WASHINGTON GEOLOGICAL SURVEY

HENRY LANDES, State Geologist

BULLETIN No. 6

GEOLOGY AND ORE DEPOSITS

OF THE

BLEWETT MINING DISTRICT

By CHARLES E. WEAVER

OLYMPIA, WASH.:
E. L. BOARDMAN, PUBLIC PRINTER
1911

LETTER OF TRANSMITTAL.

Governor M. E. Hay, Chairman, and Members of the Board of Geological Survey:

GENTLEMEN—I have the honor to submit herewith a report entitled "Geology and Ore Deposits of the Blewett Mining District," by Charles E. Weaver, with the recommendation that it be printed as Bulletin No. 6 of the Survey reports.

Very respectfully,

HENRY LANDES,
State Geologist.

University Station, Seattle, March 10, 1911.

CONTENTS.

ILLUSTRATIONS.

INTRODUCTION.

FIELD WORK AND ACKNOWLEDGMENTS.

This report includes a short history of the development of the mining industry at Blewett, the topography and physiography, a description of the geological formations, the geological history and structure, a description of the character and appearance of the ores, and a detailed description of the different mining properties.

The geological field work, of which this paper represents the results, was carried on during the summer of 1910. Work was begun on June 13th, and continued until July 19th and during this time the areal geological mapping was completed and the preliminary underground work begun. Later in the same year, during October, two weeks were spent in completing the detailed underground surveys and in studying the ore deposits. During the summer of 1908 six weeks were devoted to a study of this district but not in connection with the State Geological Survey. The results of that investigation have been freely used in connection with the data obtained during 1910. The winter months of 1910 and 1911 have been largely devoted to working up the data collected and in preparing this report.

The topographic base map, upon which the areal geology has been represented, is a photographic enlargement of a small portion of the Mount Stuart Quadrangle and is on a scale of three inches to the mile. Because of this enlargement many of the represented topographic features may appear distorted. Since the most extensive mining developments in this district have been carried on in Culver gulch, a detailed topographic map on the scale of 200 feet to the inch and having an area of about one square mile, was constructed. Upon this map every important outcrop, including the veins, was mapped by a transit and stadia survey and tied to the United States mineral monument at the head of Culver gulch.

The writer was assisted in this work by Mr. Charles R. Fettke, who did a large part of the drafting and a part of the assaying. The chemical analyses were made by Mr. M. C. Taylor, of the department of chemistry at the University of Washington, and the majority of the assays were made by Professor C. R. Corey, of the School of Mines at the State University. Many of the chemical analyses included in this report have been taken from the Mount Stuart and Snoqualmie folios of the United States Geological Survey but due credit has been given in each case.

Throughout this work much assistance has been willingly rendered by the miners and citizens of Blewett. The larger number of the old mine workings were caved and impossible of entrance but several of the miners in this camp who had worked in them during the early days of mining identified many of the openings and with the writer explored them as far as possible. The writer wishes to thank all who have assisted in the carrying out of this work and especially Mr. Jack McCarty and Mr. John Olden, the latter having been one of the pioneers in the early seventies when the first quartz locations were being made. From both of these men much valuable information concerning the early history of the district was obtained. Many thanks are due Mr. Thomas A. Parish of Seattle, President of the Alta Vista Mining Company at Blewett, for the many courtesies extended, including the placing of his cabin at the disposal of the party throughout the entire field season.

LOCATION AND AREA OF THE DISTRICT.

The Blewett Mining District, which is sometimes known as the Peshastin Mining District, is located in the south central part of Chelan county, approximately in the center of the State of Washington. It lies west of Columbia river and east of the main divide of the Cascade mountains, in the upper part of Peshastin valley, at about latitude 47° and 25′ north and longitude 120° and 40′ west. The areal limits of this district are not very well defined but it may be said in general to comprise all of the mining claims situated on upper Peshastin creek down as far as the mouth of Ingalls creek, including its tributaries. The claims on Negro creek, however, are often referred to as a part of the Negro Creek Mining District. In this report an area of approximately nine square miles has been selected which includes all the more important mining claims, with the possible exception of two or three on the upper part of Negro creek.

About one-half mile to the east of the center of this areal map at a point where Culver gulch empties into Peshastin creek, is situated the town of Blewett, at an altitude of 2,328 feet, with a population at the present time of about forty people. This town is situated on the north side of the Wenatchee mountains, which form the water shed between the drainage basin of Wenatchee river on the north and Yakima river on the south. To the north the most important towns are Wenatchee, at the junction of the Wenatchee and Columbia rivers, Cashmere, Peshastin, and Leavenworth. These towns are all connected by the main transcontinental line of the Great Northern Railway, on the Seattle-Spokane division. To the south are the towns of Cle Elum and Ellensburg, which are connected by two transcontinental railroads, the Northern Pacific, and the Chicago, Milwaukee and Puget Sound. In the early days of mining, Blewett was

Scale

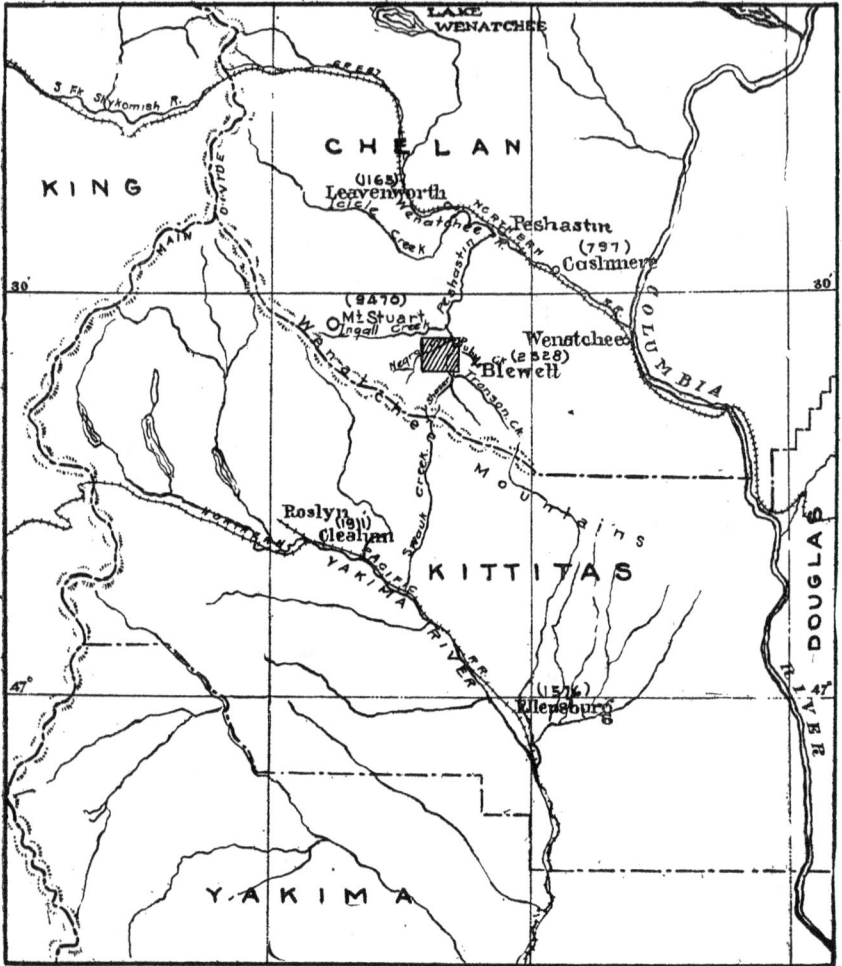

MAP OF EASTERN PORTION OF CASCADE MOUNTAINS
BETWEEN THE WENATCHEE AND YAKIMA RIVERS.

Area Representing Blewett Mining District
Areal Geology Mapped

connected with the outside world by road to Cle Elum, a distance
of thirty-two miles. This road extends from Blewett southward
up Peshastin creek for a distance of about nine miles, to the summit of the Wenatchee mountain ridge, the last mile of the nine
having a 500-foot grade. South of this divide it passes down to
the valley of Swauk creek. For the first two miles it has a 500-foot
grade, and then extends down this same creek to a point about
four miles below the town of Liberty and thence turns westward
to Cle Elum. Later, in 1898, the wagon road to the north was
completed through to the town of Peshastin, a distance of
eighteen miles, and from there is connected by both wagon road
and railway to Leavenworth and Wenatchee. The distance from
Peshastin to Seattle along the line of the Great Northern Railway is one hundred and forty-six miles, and from Spokane is one
hundred and ninety-three miles. Along the road from Peshastin
to the mouth of Ingalls creek the grade is very moderate, averaging about forty-four feet per mile. From Ingalls creek southward to the town of Blewett it increases to 100 feet to the mile,
but owing to the steep, canyon-like character of the valley above
Ingalls creek, the grade becomes exceedingly irregular and in
many places is much greater. Blewett is connected with Peshas-
in over this road by a tri-weekly stage and mail service and also
with Seattle by long distance telephone connection by way of
Cle Elum.

INDUSTRIES AND SETTLEMENTS.

The only industry of any importance within the Blewett district is mining. To the north of Blewett, in the lower portions
of Peshastin valley, there are many small farms where alfalfa,
apples and small fruits are raised. The extreme northern portion of Peshastin valley is a part of the famous Wenatchee apple
orchard district. To the south, in the valley of the Swauk, wheat

and other crops constitute the chief products. During the summer months large bands of sheep roam over the hills and mountain slopes and feed on the native grasses. In the town of Blewett there is one store, hotel, and postoffice. During the early days of mining Blewett was a thriving town with a population of several hundred people, but at the present time, due to the inactivity of mining, it has dwindled to a mere handful.

LITERATURE.

The literature bearing upon the geology and ore deposits of the Blewett Mining District and the surrounding country is not extensive, but a part of it represents the results of thorough detailed investigation by prominent geologists. The following list embraces the more important publications referring directly to the geology of this region. Many other references may be found in technical magazines and newspaper clippings which, while they refer to many phases of the history of mining of the region, are not deemed of sufficient geological interest to be included in the following list:

RUSSELL, ISRAEL C., A Geological Reconnoissance in Central Washington, Bulletin 108, U. S. Geological Survey, Washington, D. C., 1893.

This paper describes the scientific result of a general reconnoissance of the eastern portion of the Cascade mountains, from Lake Chelan southward nearly to the Oregon boundary. The region actually involved in the Blewett district was not visited but many of the same formations occurring here were studied in other parts of the region examined. The Pre-Tertiary formations are not differentiated but simply described as the "crystalline rocks" lying unconformably beneath the later sediments and lavas, and the Tertiary formations now known as Swauk and Roslyn were grouped together under the general term Kittitas system. Professor Russell says, however (page 20), "Future study may show that this system should be sub-divided but as only a small portion of the region it occupies was traversed by me the classification here proposed will be sufficient for present needs." Resting unconformably upon the Kittitas system are a series of lavas which he calls the "Columbia river lava" and above these are fresh water lake beds, which he correlates in age with the John Day beds in Oregon. All of these formations taken together form the bulk of the Cascade range, whose age he provisionally placed as Post-Miocene. No reference is made to any of the ore deposits on Peshastin creek. A list of fossil leaves collected from the Kittitas strata near Wenatchee and identified by Dr. F. H. Knowlton is given.

BETHUNE, GEORGE A., Mines and Minerals of Washington, Annual Report of George A. Bethune, First State Geologist, Olympia, Washington, 1891.

This report does not describe the geological features of the region. On page 92 he gives a very short account of the properties in the Peshastin district, without discussing the ores. Some mention is made of the early history of the camp.

RUSSELL, ISRAEL C., Preliminary Paper on the Geology of the Cascade Mountains in Northern Washington, 20th Annual Report, U. S. Geological Survey, 1898-99, Part II, pp. 83-210. Plates 8-20, Washington, D. C., 1900.

In this paper Professor Russell discusses the geology of an area aproximately sixty miles in width and extending from the Northern Pacific Railway, where it crosses the Cascade mountains, northward to the United States-Canadian boundary, a distance of about 100 miles. The Blewett district was visited and the formations studied. The granodiorite and serpentine, as well as the greenstones and slates, are described; the latter two corresponding to what are now known as the Hawkins and Peshastin formations. The Tertiary sandstones, which were originally called the Kittitas system, are now subdivided into two new formations which he designates as the Swauk and Roslyn, and which are separated by an extensive series of lava flows.

Upon palaeobotanical evidence they are both assigned to the Eocene. He considers the complex mass of the Cascade mountains to have undergone erosion during the late Tertiary times, and then to have been uplifted to an elevation of about 7,500 feet as a complex warped dome. The relations of the larger valleys to uplift are discussed in detail as well as the areal distribution and the geological work of the glaciers.

Considerable attention is given to the discussion of the ore deposits in the Blewett district, both placer and quartz. His description of these deposits can best be stated in his own words (page 208), "The country rock in the Peshastin district is mainly schist, serpentine, greenstone, etc., which has been much disturbed since the mineral bearing veins were formed. On account of the disturbances that have affected the rock and the small extent of the veins, most of which are rather 'gash veins' than true fissure veins, extensive and continuous bodies of gold-bearing quartz are scarcely to be expected. Rich quartz has been discovered, however, and at least one of the mines has paid well. A belt of rocks in which the mines at Blewett occur, extends for some eighteen or twenty miles to the westward, about the southern base of the central granitic core of the Wenatchee mountains and northward to beyond Leavenworth and has been found to carry gold at numerous localities. This same complex belt of rocks has been found also to contain copper, nickel, cobalt, and mercury (cinnabar) in many places. It is safe to say that several thousand mineral locations have been made in this region, more especially along Negro and Ingalls creeks, on the headwaters of Icicle and Fortune creeks, and in the mountains from which flow the several branches of the north and middle forks of Teanaway river. Only a few of these prospects have been opened so as to show what the conditions really are, and with the exception of a few locations on Negro creek, no actual mining operations have been undertaken. The fact that so few prospects have been developed, and the total absence of paying mines in 1898, lead to the inference that this region is not promising from the miner's point of view. Although rich ores seem to have

Town of Blewett, looking north. Situated in Peshastin Valley.

been discovered in many instances, no large bodies of such ores have as yet been revealed. To the geologist the main difficulty seems to be the many disturbances that have affected the region since the deposition of the ores. The numerous faults and the large areas where the rocks have been crushed and displaced make it evident that only the most careful exploration, guided by a critical knowledge of the geological conditions, can hope to lead to success."

HODGES, L. K., Mining in the Pacific Northwest, published by the Seatle Post-Intelligencer, pp. 70-76, on the Peshastin District, Seattle, 1897.

This paper describes the general geography of the Blewett and Negro Creek Mining Districts. Very little definite information is given concerning the geology or nature of ore deposition. Considerable attention is given to a discussion of the early attempts at mining and the development of mining operations up to the year 1897. A small map is included showing the position of the various mining claims at that time.

KNAPP, A. E., The Metaliferous Resources of Washington, Peshastin District, pp. 94-98, Annual Report, Vol. 1, Washington State Geological Survey (Landes), 1901, Olympia, Washington.

This paper notes in a general way the character of the formations and the veins included in them. The ores are said to be free milling and concentrating, and to occur as pocket veins, or enriched zones along lines of crushing. A short description is given of the mining operations carried on in each of the important claims.

SMITH, GEORGE OTIS, AND WILLIS, BAILEY; Contributions to the Geology of Washington, pp. 1-97, Professional Paper No. 19, U. S. Geological Survey, Washington, D. C., 1903.

In this paper the results of detailed areal geological work are presented. A map showing the areal distribution of the formations west of Blewett is included, and the formations which were

—2

formerly designated as the "crystalline rocks" are now sub-
divided into the Easton schist, Peshastin formation, and Hawkins formation. The Tertiary, sedimentary, and igneous rocks
are also described and their structural relations discussed. Special attention is given to the physiography and the development
of the topographic features. The conclusion is drawn that there
were two periods during Tertiary times of complex warping
of the Cascades, separated by a period of peneplanation. The
evidence supporting this conclusion is given, as well as the problems involved in the second period of warping or final uplift of
the Cascade mountains. No discussion of the ore deposits is
attempted.

SMITH, GEORGE OTIS; The Mount Stuart Folio, No. 106, Washington, U. S. Geological Survey, Geology Surveyed 1898-
99; Washington, D. C., 1904.

This folio represents the detailed areal geological mapping of
about 800 square miles and in the northern part of this area is
situated the Blewett mining district. All of the formations are
described, ranging from the earliest metamorphic rocks up to
latest Quaternary, including glacial and alluvial deposits. The
different structural features are discussed and an outline of the
geological and geographic history is given. The quartz and
placer deposits on Peshastin and Negro creeks are described,
and their origin and geological relations discussed. Especial
attention is given to a discussion of the Warrior General mine,
which he says is located (page 9) "In a zone of sheared serpentine where the mineral bearing solutions have found conditions
favorable for ore deposition. This mineral zone has a general
east-west course and extends from east of Blewett across Peshastin creek, up Culver gulch and across the valley of Negro creek.

The Warrior General vein has a trend of north 70° to 80°
east, and is very irregular in width. In the walls the serpentine
is often talc-like in appearance while the compact, white quartz
of the vein is sometimes banded with green talcose material.
Sulphides are present in the ore but are not at all prominent.
The values are mostly in free gold, which is fine, although in

some of the richer quartz the flakes may be detected with the un-aided eye." Regarding the placer deposits of the Peshastin district, Dr. Smith says: "The gravels appear to be gold-bearing throughout and the gold is rarely uniform in distribution. The largest nuggets are found on the irregular surface of the Pre-Tertiary slate which forms the bed rock * * * The largest nuggets found in the Peshastin district are less than an ounce in weight. * * * The Peshastin gold is fairly coarse and easily saved. The gold is high grade, being worth about $18.00 an ounce."

SMITH, GEORGE OTIS; Geology of the Snoqualmie Folio, Washington, No. 139, U. S. Geological Survey, Geology Surveyed 1899 and 1902, Washington, D. C., 1906.

In a general way the formations occurring in this quadrangle, which is situated directly west of the Mount Stuart quadrangle, are the same and whatever may be said in regard to the more important geological features of one applies to the other.

GEOLOGY AND ORE DEPOSITS OF THE BLEWETT MINING DISTRICT.

CHAPTER I.

PHYSIOGRAPHY.

TOPOGRAPHY.

GENERAL STATEMENT.

The topography of the area involved in the Blewett district is a part of the general topography of the eastern portion of the Cascade mountains, south of the Wenatchee valley and north of the Yakima. This region represents a broad spur of the Cascade mountains extending in a general east to northwest direction from Columbia river up through Mount Stuart and thence northwesterly over the main divide to the western slope of the mountain. One of the striking features of this broad spur is a ridge known as the Wenatchee mountains, which begins near the Columbia river, at an elevation of about 1,000 feet and gradually increases until it culminates in Mount Stuart, at an elevation of 9,470 feet, and then decreases until it reaches the main divide of the Cascades. This ridge forms the divide between the two valleys just mentioned and, up near its crest, head some of the large tributaries to Yakima and Wenatchee rivers. Among these on the north are Icicle, Ingalls, Negro, and Peshastin creeks. Throughout the larger part of this region the mountains are bold and rugged and often impassable. The ridges are narrow with steep slopes and canyon-like valleys, with occasional cascades along their courses. The lower portions af the larger valleys are wider and in many places alluvial plains have been built up. The general altitude of the peaks and ridges within this area decrease from the main divide on the west to Columbia river on the east as well as from the crest of the Wenatchee mountain ridge northward and southward to the valleys of Wenatchee and Yakima rivers.

The Blewett mining district lies about ten miles north of the Wenatchee mountain crest and twenty miles east of Mount Stuart, but within the drainage basin of Peshastin and Negro creeks. The topographic features of this region present bold, rugged appearances, not so steep or precipitous, however, as farther to the west near the main divide, nor so low and rounded as near the Columbia river. Conforming to the general easterly slope of the Cascades in the broad area just described, the general level of the higher ridges and peaks decreases from an elevation of 5,500 feet in the southwest to 5,000 feet in the eastern portion of the map. Two large valleys, which join near the north part of the area and which have numerous smaller canyons and gulches draining into them, stand out as a prominent feature of the topography. These smaller creeks and gulches are characterized by steep slopes and high grades, and occasionally present small waterfalls.

DRAINAGE.

The larger part of the drainage within the area of the Blewett district finds its way into Peshastin creek and its tributary, Negro creek. A very small area in the northwestern part of the district drains down to Ingalls creek, which in turn, about five miles below Blewett, drains into Peshastin creek, and the latter about eighteen miles below Blewett joins Wenatchee river, which a few miles below, empties into Columbia river at the city of Wenatchee.

Peshastin creek has a high grade, approximately 130 feet to the mile, within the north and south limits of the district. To the south, above Blewett, it receives the drainage from Shaser and Tronson creeks. At a point on Peshastin creek four miles north of the Wenatchee mountain divide it becomes a small mountain stream and the grade from there to Blewett is about eighty feet to the mile. From the intersection of the western boundary of the map and Negro creek to the mouth of the same creek the grade is approximately 430 feet to the mile. The smaller streams tributary to Peshastin and Negro creeks, such as Ruby creek, King creek, Culver Springs gulch, Culver gulch, and Bear creek, carry some water throughout the year and during the springtime often carry considerable volume.

No official records of the volume of water passing through Peshastin creek at Blewett have ever been made. However, according to estimates made by General J. D. McIntire, who was in charge of the La Rica property for a time, the flow during the dry season amounts to ten cubic feet per second. In the winter it is much greater.

FORMS OF THE SURFACE.

The lowest elevation recorded within this district is 2,150 feet at the north end of the map on Peshastin creek just below where it is joined by Negro creek and the highest elevation is 5,650 feet in the southwestern corner. On the high ridge between Shaser and Negro creeks, at the head of Bear creek, at the point where this maximum elevation was obtained, this same ridge swings northeasterly and becomes the divide between Negro creek on the north and Peshastin on the south and east. King creek, Culver Springs gulch, and Culver gulch head along the eastern slopes of this divide, and Bear creek along the northern. At the head of Culver gulch on this ridge is located the United States mineral monument for the Blewett-Negro creek mining district, at an elevation of 4,216 feet. The highest elevation reached on the divide between Negro and Ingalls creeks is 5,000 feet, and from the crest down to the valley floor the slopes are exceptionally steep. On the east side of Peshastin creek at Windmill Point the elevation attained is 4,250 feet, and on the high knob south of King creek 4,650 feet.

GLACIATION.

No evidence of the area involved in this district ever having been glaciated could be found. There is direct evidence, however, of glaciation in the surrounding region. Such valleys as Icicle and Ingalls show evidences of having been occupied by glaciers as well as Peshastin creek below its junction with Ingalls creek. These glaciers had their gathering ground high up near the crest of the Cascade mountains and especially in the region about Mount Stuart, and from there advanced downwards and partially filled the pre-glacial stream valleys and then finally, after their retreat, left moraines and erratic bowlders strewn along their former path.

CLIMATE.

The climate at Blewett is representative of the intermediate altitudes on the eastern slopes of the Cascades. The winter temperature is higher than at Wenatchee, Leavenworth, or Ellensburg. The summer heat, while sometimes considerable, is not so great as that at the above mentioned places, because of the influence of the colder air currents after having passed over the more elevated and sometimes snow-capped ridges to the west. The annual rainfall is higher than at either Ellensburg or Wenatchee. Snow begins falling late in November but generally disappears from the valley about Blewett by April, and from the higher ridges late in May. No official records of the temperature, rainfall, or snowfall have been kept at Blewett. The nearest points at which these records have been kept are Cle Elum, Lake Kachess, and Lake Keechelus. The following table has been prepared by Professor E. J. Saunders, of the University of Washington:

LOCATION	Elevation. Feet	Annual precipitation. Inches	Annual snow fall. Inches	Mean yearly temp. Fahr.	Mean temp. warmest month	Mean temp. coldest month
Lake Keechelus	2,479	61.90	230.7	42.4	93	16
Lake Kachess	2,235	56.09	167.0	42.4	60	20
Lake Ole Elum..................	2,171	31.24	113.0	....	....	....
Wenatchee	1,169	15.50	75.9	48	70	27
Cle Elum	1,930	26.60	79.5	45	64	28

The highest temperature ever recorded at Wenatchee was 104°; at Cle Elum, 101°. The lowest temperature ever recorded at Wenatchee was —16°; at Cle Elum, —24°; at Ellensburg, —29°. The elevation at Blewett corresponds most closely to that of Lake Kachess, but because of the greater distance from

the summit of the Cascades the mean yearly temperature probably corresponds more closely to that of Cle Elum; the annual
rainfall is probably intermediate between that of Wenatchee and
Cle Elum; the annual snowfall perhaps as great as that of Lake
Cle Elum.

VEGETATION.

Originally this entire district was covered with timber consisting chiefly of the following varieties: the tamarack, the yellow
pine and the red fir.   These forests are open in character, except on the lower slopes and about the stream bottoms, which are
generally heavily covered with thickets and small shrubs with
some of the larger trees scattered among them.   In Peshastin
valley and up Culver gulch much timber has been removed for
local building purposes and for mine timbering.   Above the altitudes of 5,000 feet the trees become smaller but not so small as
the stunted forms on the much higher altitudes outside of
the map.

RELATION OF THE PRESENT TOPOGRAPHY TO THE GENERAL GEOLOGY.

The explanation of the present topography and the various
relations of the streams and valleys to each other is in a large
part to be explained from a study of the geological conditions in
a much larger portion of the Cascade mountains, of which the
Blewett district is only a very small fragment.   Fortunately,
geologists with a much wider experience have studied this larger
area in detail and have been enabled to arrive at some definite
conclusions.   A discussion of the geological events concerned in
producing these changes will not be considered here, except in
so far as they relate to the principal drainage features within
the area involved.   As has been stated before, one of the most
important factors connected with the final uplift of the Cascade
mountain mass was the differential warping of the uplifted dome,
which in this particular portion of the Cascades resulted in the
formation of the long ridge now known as the Wenatchee mountains.   Immediately upon the initiation of the upward movement drainage lines began to adapt themselves to the new con-

ditions and as a result such streams as Ingalls creek and Peshastin creek developed. As the Wenatchee mountain mass kept rising the grade of the newly developed streams was increased and they soon encountered in their downward cutting rock of differential resistance. As a result the small lateral streams began to excavate their channels in those rocks which would offer the least resistance, giving rise to such streams as Negro creek, Ruby creek, Shaser creek, and Tronson creek. As time passed on this method of development increased until the present topographical features were attained. Outside of the area of this map many of the topographic features due to this cause have been modified by the action of the glaciers. Within the area of this district they have played no direct part in developing the present topography.

CHAPTER II.
GENERAL GEOLOGY.

INTRODUCTION.

The general geology of the Blewett mining district is but a small fragment of the geology of the eastern portion of the Cascade mountains. Because of this fact, the elucidation of nearly all of the intricate geologic problems involved in this district must be partially sought for in the broader field surrounding this district. A part of this broader field has been studied and areally mapped and published in the Mount Stuart and Snoqualmie geological folios. Hence all of the formations occurring in the Blewett district have been previously described, and in the following discussion an attempt will be made to represent in as much detail as possible the more salient features of each formation and their relation to the broader field outside, and the problems involved in the deposition of the ore. Many of the formations occurring outside of the district do not occur within it. The oldest formation found in this district consists of a lower series of quartzites, slates, quartzite-conglomerates and cherts, overlaid apparently unconformably by an upper series of brecciated lavas and tuffs, together with massive igneous rocks. No fossiliferous evidence has been obtained to determine their age, but from their close resemblance to formations in the north,* whose age is known, they are provisionally assigned to the Carboniferous. After their deposition they underwent deformation resulting in their metamorphism. Accompanying this deformation were great intrusions of peridotite and other associated basic igneous rocks which were later more or less altered to serpentine. After this,

---

*Ann. Report, Geological Survey, Canada, New Series, Vol. 7, 1894, pp. 37 B-49-B.

erosion became active and deformational movements produced a warping, allowing great drainage basins to form, and in these basins were accumulated sandstones and shales belonging to the Swauk formation and attaining a maximum thickness of perhaps 5,000 feet. Cutting through the Swauk sandstone are numerous diabasic dikes leading up to and connecting with surface lava flows of the Teanaway basalt. Also there are intruded into the Swauk and the older rocks as well, plutonic masses and sills of gabbro which were probably injected either just before or contemporaneous with the diabase. Overlying these are later sedimentary lake deposits known as the Roslyn formation. Above these are still younger sediments and lavas and tuffs but no rocks younger than the diabase occur within the Blewett district proper with the exception of Quaternary gravels and alluvium.

CONTACT SCHIST.

The rock referred to in this report as a contact schist has been recognized at but one locality in this district. About one mile south of Blewett on the west side of Peshastin creek at Kendall creek it outcrops near the road. A few hundred feet above this outcrop where the county road crosses Peshastin creek a long narrow dike of granodiorite, trending nearly east and west, occurs. On both sides of Kendall creek and in the creek itself this schist stands out prominently. On the south it lies in contact with the granodiorite, and on the north with the peridotite. In general appearance it is distinctly banded and filled with small gash seams of quartz. Each of the individual bands or lamellae ranges from one-fourth to one-sixteenth of an inch in thickness, and consist of fine grained, gray chert-like appearing material, alternating with white quartz. Occasionally bands of mica-schist occur. As a rule the bands are twisted and contorted but sometimes are nearly straight with a persistent strike and dip. Observations were made on the bluff on the north side of Kendall creek, where the strike was found to be north 20° west, and the dip about 75° to the southwest. On the south side of the same creek and at the creek level, a small tunnel has been driven in along the strike of the formation for a distance of thirty feet. In the tunnel the quartz bands predominate. They are not persistent in character and are rarely ever more than two inches in thickness and pinch out both above and below. In addition numerous stringers of quartz of the same general character as those occurring in bands cut obliquely across the strike and dip of the formation. No economic value attaches to the quartz on account of the prevailingly low assays obtained. The total area of the entire outcrop is not over 5,000 square feet.

While this individual outcrop does not appear on the geologic map of the Mount Stuart Folio, yet similar rocks are mapped and described near the contact of the Mount Stuart granodiorite massif with the serpentine, in the vicinity of Mount Stuart. In that region Dr. Smith found apophyses of the granodiorite and also of the serpentine extending into the schist and hence concluded that it represented some formation older than the serpentine which had been metamorphosed possibly by both. In the outcrop described in this report the serpentine is clearly intrusive in the schist. The main mass of the granodiorite dike is also included but no small apophyses were seen extending from it into the schist, although they may occur and have escaped detection. No definite age can be assigned to this rock, except that it is older than the peridotite or serpentine. It may represent an extremely metamorphosed phase of the Peshastin or Hawkins formations or it may possibly represent still older rocks. It is apparent, however, that this extreme phase of metamorphism has been produced by the granodiorite more than by the peridotite as the contortion and twisting becomes more pronounced the nearer the contact with granodiorite.

This rock is described first among the geological formations represented and is placed at the foot of the geological column in the legend on the areal geological map, but that does not necessarily indicate that it is positively older than the Peshastin or Hawkins formations. It has been designated the "Contact

Schist" because rocks* similar in character and geologic relations have already been described as such by Dr. Smith in the Mount Stuart Folio.

PESHASTIN FORMATION.

AREAL DISTRIBUTION.—When not covered over by later sediments, the series of rocks grouped together under the general term Peshastin formation form an important part of the areal geology in that portion of the Cascades near Mount Stuart. Outcrops of rock similar in character occur in various parts of the Northern Cascade mountains, which may be a part of the same general formation described here, but at the present time there is no direct evidence to make a definite statement. In the Mount Stuart and Snoqualmie quadrangles, where the geology has been carefully mapped, the areal distribution of the rocks belonging to this formation is definitely known. Surrounding the southern part of the Mount Stuart massif of granodiorite there is a fringe of igneous and metamorphic rock ranging from three to ten miles in width and extending from Peshastin creek on the east, westward along the Chelan-Kittitas county boundary for a distance of about twenty miles. This belt is overlain to the south and east by later sediments belonging to the Swauk formation. Within this belt the outcrops of the Peshastin formation occur as irregular shaped patches enclosed in serpentine and gabbro.

Within the Blewett district proper the Peshastin formation occupies a prominent position on the areal geological map. Three separate outcrops are represented: The largest has an area of about two square miles and lies to the north of Blewett on both sides of Peshastin creek and over the ridge on Negro creek and from there continues to the north boundary of the map. It outcrops on both sides of Negro creek to a point about one and one-fourth miles from its mouth and then extends westward as a narrow belt about 1,500 feet wide along the south side of the ridge between Negro and Ingalls creeks to the

---

*Geology of Mt. Stuart Folio, p. 5, U. S. Geological Survey, 1904, Washington, D. C.

western limits of the map. Two other very small outcrops occur, the larger of which is situated in the southeast corner of
the map on the south side of Sheep mountain as an elliptical
shaped mass, apparently pitching downwards at a very low angle
beneath the Hawkins formation. The second patch occurs at
the head of the south fork of King creek.

GENERAL DESCRIPTION.—The rocks composing the Peshastin formation are prevailingly black slates and fine grained, dark
colored quartzites. There are subordinate proportions of black
grit, ranging to a fine conglomerate, together with occasional
narrow bands of green and red chert and ferruginous slate. This
entire series has undergone more or less extensive metamorphism,
but the original character of the sediments is in most places
plainly discernible. On the east side of Peshastin creek about
one-fourth mile below Culver gulch, a quartzitic conglomerate
belt occurs having an average thickness of nearly 100 feet, but
owing to the deformation which it has undergone it is more or less
crushed and faulted, and in places neither the upper nor lower
limits could be traced. Originally this rock was a fine grained
conglomerate, grading into a coarse sandy grit. The pebbles
contained in it have in places been more or less drawn out ellip-
tically, but often retain their original shape. They range in size
from ordinary sand grains to one-half inch in diameter. These
pebbles consist of blue, white and dark gray quartz, probably
representing grains of a former quartzite, of red and brown
chert, of black volcanic glass, of dense, hard compact greenstone
and occasionally of much altered white to light gray material
which originally may have been volcanic tuff. In some of the hand
specimens water worn pebbles of black clay slate were observed.
The whole mass contains more or less secondary pyrite impregnated through it. Many of the pebbles, especially of the quartz
and chert, are noticeably cracked and fissured and filled with
secondary quartz. This fracturing apparently occurred prior
to the formation of the pebbles, while they were still a part of
some pre-Peshastin massive rock.   The conglomerate masses
have been metamorphosed along with the other members of the

Peshastin formation, and in consequence have suffered considerable shattering. These quartzitic conglomerates also occur interbedded with slate and quartzites as may be observed on the north side of Negro creek, about two miles above its mouth.

The clay slates are extremely fine grained, ranging in color from a bluish-gray or gray to black, and have a close, crystalline texture. They tend to cleave in parallel plates of varying thickness but the size of the plates is very irregular, due to numerous but less prominent joint planes which intersect the slate mass, in various directions. Often the slates change in character, indicating changes in sedimentation during their deposition and prior to their metamorphism. Commonly bands of various thicknesses of fine grained quartzite alternate with the slate and grade into each other, showing distinctly the original stratification bedding plane. Often the clay slates appear more massive in character, and instead of being characterized by cleavage planes, possess a fissile character, become coarser grained, and break with a conchoidal fracture. The zones of most prominent shearing are approximately parallel to the stratification or bedding plane. Near the mouth of Negro creek and along the county road from this point to Blewett excellent exposures of the strike and dip, cleavage and joint planes may be obtained. The average of about thirty different observations taken on the strike and dip were—strike, N. 5° W. and dip, 80° to 85° to the southwest.

HAWKINS FORMATION.

AREAL DISTRIBUTION.—The rocks which are described as the Hawkins formation constitute a prominent feature of the areal geology within the Blewett district proper, and are a part of a much broader belt extending east and west from Mount Stuart. They have been areally mapped in the Mount Stuart and Snoqualmie quadrangles, where they may be seen to extend as more or less disconnected patches from Peshastin creek on the east, westward to Mount Hawkins, and thence northwesterly. In all of these localities they are intimately associated with the Peshastin formation and with a peridotite intrusive mass and undoubt-edly would have a much wider areal distribution if the later over-lying Tertiary formations were removed.

Within the areal geological map accompanying this report the Hawkins formation is well represented. In the western portion of this district three separate areas have been mapped. One of these has an area of about one square mile and lies partly on the north and partly on the south of Negro creek. In the southwestern corner there is an area of about two-thirds of one square mile, and just east of this another area about 1,500 feet in width and one mile in length. On the east side of Pe-shastin creek the Hawkins formation occupies an area of about one-quarter of a square mile, and includes the rocks forming Windmill point. South of this an area averaging 600 feet in width and one mile in length crosses over to the west side of Pe-shastin creek and extends up to Sheep mountain. In addition to those already mentioned, several outcrops occurring in the serpentine have been mapped, and numerous patches too small to represent occur, but have not been separated from the ser-pentine.

GENERAL DESCRIPTION.—The rocks grouped under the gen-eral term Hawkins formation constitute the bolder crags and pinnacles such as Sheep mountain, Iron mountain, and the high ragged ridge between Culver gulch and Negro creek. The hard and compact texture of the materials composing these rocks have resisted weathering compared with the more easily eroded ad-jacent rocks. These rocks when viewed from a distance have a black, basaltic appearance and as a rule are devoid of much soil. On the north and upper side of Culver gulch this formation out-crops prominently with a broken surface and sharp crag-like crest. Below it outcrops the serpentines, in which are in-cluded some of the richest veins of ore in the district. These veins extend over into the Hawkins but as far as can be deter-mined the ore values decrease. The rocks composing this for-mation are represented by volcanic breccias, tuffs, and inter-calated volcanic flows. These are nearly everywhere intricately

A—General view of district, looking west from the hills east of Peshastin creek. Culver gulch in foreground. Negro creek valley immediately back and Ingalls creek valley in the rear.

B—North side of Culver gulch, showing outcrops of "Nickel Dike."

mingled, as may be seen on the eastern side of Peshastin creek, and on the slopes leading up to Windmill point. In the exposures here represented great masses of rock composed of angular and partially worn fragments occur. Among these bowlders are gabbro, diabase, diorite and andesite porphyry; and irregular shaped fragments of scoriaceous material and volcanic tuff. These are partially cemented by silica and are intersected by secondary veins of quartz and calcite. In some places the fragments composing this mass are made up of quartzites, slates and metamorphic grits, and closely resembling those in the Peshastin formation, occurring on Peshastin creek just north of Blewett. In still other places fragments of greenstone occur intermixed with the other material. The size of these angular fragments ranges from small grains up to twenty or more feet in diameter, the larger bowlders being for the most part composed of gabbro porphyry nearly free from alteration and apparently genetically associated with the basic flows.

The interbedded flows are generally broken and in places tuffaceous and have scattered through them fragments of the rocks just mentioned. When observed from a distance, these rocks appear black, but on closer examination they are seen to be a dark green. Near the contact with the serpentine they are often white, due to the coating and impregnations of magnesium carbonate, which has been derived by solution from the serpentine mass. An instance of this may be seen in the mass leading up to Sheep mountain. On the sheer walls forming the crag of this mountain these rocks are deeply stained a bluish-green and in places contain small specks of chalcopyrite. The fragments composing this mass when examined are seen to be composed of brecciated fragments of igneous lava, tuffs, and cherts.

PETROGRAPHY.—An examination of about thirty hand specimens collected from different outcrops of the Hawkins formation, occurring within this district, shows many variations in the character of the material composing this series of rocks. The ma-

—3

jority of the specimens are found to be metamorphosed diabasic rocks, and their corresponding lava equivalents. A typical instance of this diabasic phase is well represented by specimen No. 1143. To the naked eye it appears to be a massive, heavy, gray rock of medium grain and resembles very closely some of the gabbro phases of the peridotite mass, in fact they may be in part small apophyses of the latter injected into the Hawkins formation. They are made up of about equal amounts of feldspar and hornblende in small, allotriomorphic crystals averaging about one millimeter in size. Under the microscope these feldspars prove to be a much altered plagioclase having the chemical composition of bytownite. The hornblende is dark brown and altered to serpentine and chlorite. A very small amount of nearly colorless augite is present.

An examination of several specimens taken from the brecciated lava flows presents a fine grained, gray rock with many conspicuous, white rounded crystals embedded in it, ranging in size from one to three millimeters. Under the microscope the slide is seen to be composed of an intricate mixture of secondary alteration products, consisting of quartz, mica, calcite, and chlorite, with occasional portions of the original unaltered feldspar crystals. These alteration products represent what originally were augite, feldspar, and some orthorhombic pyroxene. Remnants of augite and plagioclase sometimes occur not very much altered. The large crystals of plagioclase are, when relatively fresh, distinctly zoned, and show albite twinning. Extinction angles were measured on one or two of these crystals yielding values of 37° and 39°. The evidence points to these rocks having been originally basic andesite and basalt flows. These two types constitute the bulk of the igneous material composing this formation.

PERIDOTITE.

AREAL DISTRIBUTION.—Ultra-basic intrusive rocks are well represented as a part of the areal geology in the eastern part of the Cascade mountains. In the Mount Stuart and Snoqualmie folios they constitute the surface rocks over an area of about 100 square miles and lie in a nearly east to west belt, averaging from two to eight miles in width and about twenty miles in length. This belt is situated south of Ingalls creek and north of the ridge known as the Wenatchee mountains. East of Peshastin creek it passes beneath the sedimentary and volcanic rocks; to the west it passes around the southern side of Mount Stuart and through Mount Hawkins, and then swings northwesterly through the Snoqualmie quadrangle into the Skykomish, where it is known to have a wide distribution but has never been areally mapped. Other known occurrences are scattered here and there through the Cascade mountains, and are probably closely related in age. Within the boundaries of the Blewett district, the peridotite, with its resultant alteration product of serpentine, has a larger areal distribution than the rocks of any other formation, approximating three-fifths of the total area. These occur in the south half of the map, with several large blocks of older rock incorporated into them, as well as later intrusive dikes. In the north half of the area the peridotite occurs as a long narrow band which has been forced up into the Hawkins and Peshastin formations.

GENERAL DESCRIPTION.—The formation described in this report as peridotite covers an assemblage of ultra-basic plutonic rocks and their alteration products which occur in various degrees of serpentinization. These rocks vary extremely in character and general appearance, and in some places stand out as conspicuous topographic features. More commonly they form the less rugged ridges, because of the less resistance to erosion. This may be seen by noting the topography on the east side of Peshastin creek, immediately east of the town of Blewett. The lower 1,000 feet—while comparatively steep, is composed of soil, and has rounded, smooth surfaces, while above the 3,500-foot contour the outcrops of the Hawkins formation stand out as bare jagged rocks. The same is true on Sheep mountain when contrasted with the less rugged topography just to the north. On the west side of Peshastin creek, in the vicinity of King creek, Culver Springs gulch, and Culver gulch, the same observations may be made. Iron mountain, at the heads of Bear creek and King creek, stands out as a prominent crag and just west of it on the same ridge lies a belt of serpentine about 1,000 feet wide, forming a saddle or depression, and immediately west of this on the same ridge the topography again becomes rough and broken. Both Iron mountain on the one side of the saddle, and the jagged peaks upon the other, are composed of hardened tuffs and breccias belonging to the Hawkins formation, while the more easily eroded portion is confined to the serpentine in the saddle. On the north side of Negro creek a narrow band of serpentine again outcrops, averaging 500 feet in width and lies along the contact between the gabbro on the north and the slates and quartzites of the Peshastin formation on the south. While the area involved is very small, yet the contrast in weathering can be clearly observed. The conclusion to be drawn is that the peridotite as a rule disintegrates under weathering much easier than does the adjacent formations, and the reasons for this must lie in the physical and chemical character of the rock masses.

The physical and chemical characteristics of the rocks grouped under the general term peridotite present many variations. These apply to the general appearance, color, texture, specific gravity, lustre, fracture, hardness and mineral composition. The effects of weathering upon the peridotite differ with variations in the above mentioned characteristics. The heavy, massive, nearly unaltered varieties are least affected and constitute the more rugged features within the area mapped as peridotite. The softer altered phases of serpentine masses disintegrate more easily, leaving the surface covered with talus material, composed of fragments of serpentine having a somewhat schistose or shaly appearance, with conchoidal fracture and smooth, slicken-sided surfaces. Often the character of the rock changes every few feet, which is due to differences in the texture or character of the minerals composing the original rock, and the resulting variations in the degree of serpentinization. Occasionally the nearly fresh peridotite outcrops over considerable areas, with a massive, dark greenish, porphyritic appearance. Large crystals of bastite, the alteration product of enstatite, are easily distinguishable in a dark, fine grained ground mass. . Scattered through such masses often occur irregular shaped patches of variously colored serpentine, sometimes black, and less commonly a pure white, but prevailingly a greenish tint. Often a black, glassy variety may be seen containing small lenses of pure white serpentine, ranging in size from a few cubic inches to several thousand feet. In many cases the contact line between the two are sharp and distinct, but between the serpentine and peridotite there is generally a more gradual gradation into one another. Observations show the black glassy varieties to occur near the contact with rocks of other formations, while the greenish varieties are scattered at large through the peridotite mass. Again, the strictly serpentine phases may extend over large areas and scattered about in them occur irregular bunches of only partially serpentized peridotite. In a great many cases rocks resembling a basic diorite or gabbro porphyry occur intimately associated with the serpentine which also are partially serpentinized. The boundaries of such outcrops are indistinct. They occur in small areas and grade into the main body of the serpentine or peridotite so that it is impracticable to represent these phases on the areal map as distinct from the true peridotite. Examples of these occurrences may be seen in the divide between Culver Springs gulch and Culver gulch, at an elevation of 200 feet above Peshastin creek, and also at an elevation of twenty feet above the same creek along the water flume near Culver gulch. This belt is about 200 feet in width, and extends up the hill for a distance of 200 feet and then pinches out as a wedge. A similar rock outcrops in several places on the south side of the ridge, between Culver gulch and the Draw canyon, and also in that canyon on its north wall. These outcrops lie in the serpentine and seem to grade into it. It is possible that they may represent more acid phases of the ultra-basic magmas, such as gabbros which were intruded into the peridotite mass prior to its serpentinization, and which later were acted upon, along with the peridotite, by the same heated, upward-travelling solutions which serpentinized them. These will be discussed in the petrographical description as the Draw canyon gabbro porphyries. Because of the uncertainty of separating these from the peridotite, and because of the fact that they were apparently differentiations from the same ultra-basic magma, and intruded at approximately the same time, and later serpentinized with the peridotite, it has seemed best to include and describe all under the general term peridotite.

In many cases the line of contact upon the areal map between the peridotite and other formations has been arbitrarily drawn. This is partly because of the resemblance of the peridotite to certain phases of the igneous portions of the Hawkins formation, partly because of certain isolated patches of serpentine within the Hawkins and Peshastin, too small to represent upon the areal map, partly because small blocks of the Hawkins have been incorporated within the serpentine and extremely altered, and partly because of the irregular contact lines, due to the irregular intrusions of the peridotite and the extension of long apophyses of the latter into the Hawkins and Peshastin formations. Often alteration of certain phases of the Peshastin have occurred, due to these basic intrusions, and in several cases slide material has entirely covered the contact. On the east side of Peshastin creek, just east of Blewett, the contact between the serpentine and the Peshastin formations extends up the hillside in a general southeast direction. In many places huge blocks of the metamorphosed Peshastin grits project up into the serpentine, but the contact can be fairly well traced until the Hawkins is reached. Much of the Hawkins below Windmill point is composed of a dark, basic, igneous rock, resembling very closely some of the less altered phases of the peridotite and project up into the latter as isolated blocks which were disconnected from their original position and incorporated into the peridotite at the time of the latter's intrusion. The same is true on the south side of Negro creek, just below the United States mineral monument, along the serpentine-Hawkins contact.

In other localities the contact is sharp and knifeblade-like, as may be observed on the west side of Iron mountain before mentioned. Here the contact is nearly vertical, the Hawkins presenting a vertical cliff, while at its base the serpentine abuts up against it so that a person can stand with a foot on either side of the contact. Throughout the entire serpentine mass many small patches of rock were seen, resembling the Hawkins very closely. Some of these were not over fifty square feet in area. They are apparently blocks of the Hawkins plucked off from the roof of the intrusive mass at the time of the intrusion and caught up and scattered about within it. The smaller outcrops are not mapped as the Hawkins but are included with the peridotite. The large masses such as Sheep mountain and the one extending from the United States mineral monument to Iron mountain are probably included within the serpentine in the same manner as the smaller blocks. In some instances small apophyses of peridotite were injected up into the Hawkins and Peshastin through small fissures, extending up from the roof of the peridotite magma and later disconnected from the main mass of the magma, and are now represented as isolated lenses of serpentine. Instances of this may be observed on the north side of Culver gulch in the Peshastin formation above the Golden Eagle tunnel and on the east side of Peshastin creek, about one mile north of Blewett, and also at the head of Culver Springs gulch, just above the three cabins. The peridotite is cut along with the older and later rocks by granodiorite porphyry and diabase

Special importance attaches to the peridotite from an economic standpoint, as it is mainly within that formation that the majority of the gold-bearing veins are situated. Their relation to the serpentine will be discussed later under the chapter on Economic Geology.

PETROGRAPHICAL DESCRIPTION.—The rocks which are described as peridotites vary more in character than any other formation represented within this district, due chiefly to variations in the chemical character of the magma at the time of the intrusion. Three principal types can be described, namely:

saxonite, pyroxenite, and an extremely basic phase of gabbro. All of these are differentiations from the same igneous magma and have been more or less altered to serpentine.

Saxonite is a variety of peridotite composed primarily of olivine and enstatite. Olivine predominates, composing three-fifths of the mass. No specimens were found in which these minerals were not partially altered to a network of fibrous serpentine and iron oxide. The alteration is seen to have started around the edges of the crystals and to have extended into it along the cracks, where the serpentine fibers occur normal to the long direction of the fracture. The crystals are allotriomorphic and sometimes over five millimeters in diameter. They are nearly colorless, but often show a greenish tint, but never pleochroism. Enstatite, $(MgSiO_3)$ occurs sometimes fresh but generally altered to a variety of serpentine known as bastite. These fibers lie parallel to the prismatic cleavage which on basal sections show intersecting angles of nearly 90°. The fractures traverse the sections at all angles. The color ranges from a yellowish to olive green and shows very weak pleochroism. In addition to these two minerals, small amounts of hornblende and augite occur, forming less than five per cent of the whole rock.

Pyroxenite occurs in many places and especially at the head of Culver gulch where the basic rocks show a decrease in the amount of enstatite and olivine and a relative increase in diallage. The diallage is recognized by the occurrence on basal sections of orthopinacoidal cleavage lines, has a dark green color, and occurs in good sized, subhedral crystals partly altered to serpentine. In one crystal chlorite was recognized as a decomposition product indicating the presence of aluminum. Augite is present in small amounts and shows the typical prismatic cleavage. Two thin sections showed crystals cut nearly parallel to the clino-pinacoid, giving extinction angles of 54° and 56°. They are nearly colorless and show no pleochroism. In size they average much smaller than the diallage.

On the ridge between Draw canyon and Culver gulch several

A—General view of mine workings in the upper part of Culver gulch.
Diabase dike in the center.

B—Culver Springs gulch with arrastre in foreground, looking west. Iron
mountain composed of Hawkins breccias in upper right-hand corner.

specimens of rock were collected which showed to the naked eye very small, white crystals of feldspar, with hornblende and some augite. When examined under the microscope they proved to be basic plagioclase with a composition of about six parts calcium to one of sodium, which places them in the anorthite division of the plagioclase group. Very few of these crystals range over one millimeter in diameter, but they usually show well developed albite twinning, giving symmetrical extinction angles of 38° on sections cut nearly normal to the twinning lamellae. Brown hornblende in subhedral crystals occurs showing pronounced pleochroism and of about the same size as the feldspar. Subordinate amounts of olivine and augite together with magnetite are scattered through the section. Some of these outcrops grade into the strictly pyroxenite and peridotite phases of the mass and are serpentinized along with them. The evidence points to their having been intruded at about the same time. However, other outcrops such as the one occurring along the water flume just south of the mouth of Culver gulch, where the planes of contact with the adjacent peridotite are more sharply defined may possibly have been intruded at a later date along with the Eocene gabbro intrusions.

"NICKEL DIKES."

AREAL DISTRIBUTION AND GENERAL DESCRIPTION.—One of the characteristic features of the serpentine belt extending in an east-west direction, south of Mount Stuart, is a series of bright red belts varying in width from 100 to 500 feet. These have been described by Dr. George Otis Smith* who considered them to represent lenses of limestone detached from the Peshastin formation and caught up in the serpentine.

Within the Blewett district certain red bands extend east and west through the serpentine and Hawkins breccias but when examined below the surface are found to be serpentine. Other belts such as the one mapped on the north side of Culver gulch extending down to Negro creek, have a different appearance.

---

*Mount Stuart Geological Folio, pp. 3 and 4, U. S. Geological Survey, 1904, Washington, D. C.

When examined with the naked eye they seem to be hard, compact, fine grained, reddish colored rocks, containing small, dark, porphyritic crystals of pyroxene, averaging about 1 mm. in size.   This phase of the rock, when examined under the microscope, is found to be composed largely of a fine grained ground mass made up of interlocking grains of silica and dolomite, and small, irregular, secondary seams of opal.   In some places small crystals of siderite are scattered through the mass.   The large porphyritic crystals are chiefly enstatite more or less altered. The character of this rock changes from point to point, occasionally grading into almost pure dolomite.

The origin of this rock cannot definitely be stated, but from its close proximity to the Peshastin formation it appears to represent a portion of that formation composed largely of lime which has been extremely altered by the intrusion of the peridotite which is rich in magnesia.   The original limestone may have been dolomitized by the addition of magnesium from the serpentine.   The porphyritic phases of this rock are undoubtedly igneous in origin and represent a portion of the original peridotite magma intruded as dikes.

The chemical analysis of this rock was made by Mr. M. C. Taylor and is here inserted:

SiO_2 .... 43.22%
Al_2O_3 .... 1.40
Fe_2O_3 .... 3.20
FeO 2.10
MgO 8.40
CaO 5.44
Na_2O }
K_2O } 0.36
H_2O 0.92
    (above 110°)
CO_2 .... 36.37
                                                  _____
                                                  100.51%

GRANODIORITE.

AREAL DISTRIBUTION.—The areal geology of the region to the northwest of Blewett is characterized by vast outcrops of granodiorite representing the uncovered portion of a great underlying batholith.   The western portion of this area forms the Mount Stuart mass and is generally referred to as the Mount

Stuart massif. Surrounding this massif is a belt of rocks belonging to the Hawkins and Peshastin formations, together with serpentines, and cutting through these metamorphic rocks are small apophyses or dikes of granodiorite. Surrounding this belt of older rocks and overlying them are sandstones of the Swauk formation. The Blewett district lies in the middle zone of metamorphics and serpentine, and the granodiorite occurring within it is represented by small, irregular shaped porphyritic dikes, occurring for the most part south of King creek. The larger of these outcrops appear at the surface as a network of connecting dikes, lying on both sides of Peshastin creek and south of King creek, forming a large part of the ridge between that creek and Shaser creek. Extending from these dikes on the east side of Peshastin creek near the south boundary of the district are three small apophyses cutting through the serpentine towards Sheep mountain. A third dike about 300 feet in width and 1,500 feet in length crosses Peshastin creek between King creek and Culver Springs gulch. Just east of this outcrop, a long narrow dike of the same material, extending nearly east and west, outcrops in the serpentine between Sheep mountain on the south and the Hawkins formation on the north. The most northerly outcrop of granodiorite occurs as a long northwest to southeast dike lying on the north slope of Culver Springs gulch. Three other prominent dikes, extending in a general north to south direction, outcrop in the serpentine on the ridge between Iron mountain and the bold rugged pinnacles of the Hawkins formation on the west. In many other places throughout the serpentine mass in this area, small quartz-like appearing outcrops occur, which at times seem to be extremely acid phases of granodiorite dikes but so small that they have not been mapped.

GENERAL DESCRIPTION.—The rocks composing these granodiorites dikes vary considerably in appearance. The granodiorite from the Mount Stuart region proper is distinctly plutonic and holocrystalline in character, while the dikes outcropping in the Blewett district range from granodiorite porphyries to andesite porphyries. As a rule the wider the dike the more closely it approaches the plutonic type. The narrow dikes show the effects of more rapid cooling, and generally present well formed, subhedral to euhedral crystals embedded in a fine grained ground mass. Occasionally the feldspar and hornblende crystals have a general elongation in one direction, giving the rock the appearance of lava flows, exhibiting flow structure. The crystals in the larger mass are often of considerable size and subhedral to anhedral, but embedded in an allotriomorphic granular ground mass composed for the most part of quartz and plagioclase, with a very small amount of hornblende. The rock is prevailingly a light gray, with occasionally a bluish-green tint. It is harder and more resistant to erosion than the serpentine and for this reason stands out in contrast to the surrounding rock.

PETROGRAPHICAL DESCRIPTION.—Upon examination with the naked eye these rocks are seen to be composed chiefly of plagioclase, orthoclase, hornblende, and quartz. The light colored minerals predominate. In some of the specimens a very small amount of hornblende is present. In others it makes up at least 30% of the rock. Three distinct phases of this rock occur within the Blewett district, and a separate description of each follows:

Specimen No. 1100.—This rock was obtained from the granodiorite outcrop on the highest point along the ridge between King and Shaser creeks, at an elevation of 4,150 feet. Microscopically, this is a dense, hard, compact rock composed of white crystals of plagioclase and orthoclase, some hornblende and a very little quartz embedded in a fine grained, dark gray ground mass, and has a decidedly porphyritic appearance. The hornblende and feldspar occur in about equal proportions.

When examined under the microscope, there are found in addition to the minerals already described, a small amount of apatite, titanite, and zircon. The ground mass in which all of these crystals are embedded is made up of numerous, very small interlocking crystals of quartz and feldspar, the latter being composed of plagioclase and orthoclase in about equal proportions. Of the larger porphyritic minerals, plagioclase is the most important. It occurs in idiomorphic crystals, ranging in size from 0.3 mm. to 3.0 mm. The crystals are well formed, generally tabular in habit and show well developed albite twinning parallel to the brachypinacoid. Sections which were cut nearly normal to the albite twinning lamellae gave symmetrical extinction angles as high as 27°. These, however, were attained near the center or core of the crystal and as nearly all of these crystals exhibit pronounced zoning or banding, the angles attained near the margin average 18° to 20°. In some of the smaller crystals in which no zoning or banding is present, extinction angles of 15° occur indicating the more acid phases of labradorite, which is a calcium sodium aluminum silicate, where the ratio of calcium to sodium is nearly equal.

Fairly large crystals of green hornblende are abundant. Their shape is generally irregular. In size they range from small, fragmentary crystals less than 1 mm. up to 1.5 mm. and show the usual pronounced pleochroism. They are commonly altered to iron oxide and chlorite.

Orthoclase is found evenly distributed through the rock, but always subordinate to plagioclase. It makes up approximately 10% of the rock and occurs in crystals ranging in size from 0.2 mm. to 3.0 mm. The larger proportion of this mineral, however, is found in the fine grained ground mass.

A few well formed crystals of biotite were observed under the microscope, less than 0.5 mm. in diameter and parallel to the basal cleavage plane. An occasional prism of zircon or apatite was noted. These larger idiomorphic crystals are scattered about embedded in a ground mass made up of very small irregular grains of orthoclase and quartz, which were the last minerals to separate out from the magma.

A second type is represented by a specimen collected from the granodiorite dike which crosses Peshastin creek between Culver Springs gulch and King creek. This differs from the specimen just described in the presence of a larger number of rounded quartz crystals embedded in the finer grained ground mass. The crystals range in size from 0.3 mm. to 2.0 mm. and are generally clear and filled with small, isotropic intrusions. Hornblende is less conspicuous and the feldspars are represented by a larger percentage of orthoclase, though still subordinate to the plagioclase.

The third phase is represented by specimen No. 51-B, which was collected from one of the extremely narrow dikes cutting into the serpentine and Peshastin formations on the south side of Sheep mountain. This specimen is very fine grained, and to the naked eye seems to be made up of feldspar, quartz, and hornblende. Under the microscope the feldspar crystals are found to be mainly andesine, and orthoclase is more common than in the other two specimens just described. Hornblende is more abundant than in the case of the second specimen but not so much as in the first. Very few crystals range over 1.00 mm. in size. They are all embedded in an extremely fine ground mass, composed of a mosaic of quartz and orthoclase. A chemical analysis of this specimen has been made and is inserted in the following table, along with three analyses made by the U. S. Geological Survey and published in the Mount Stuart Folio:*

ANALYSES OF MOUNT STUART GRANODIORITE.

	I Per cent.	II Per cent.	III Per cent.	IV Per cent.
SiO_2	64.04	63.87	63.73	66.40
Al_2O_3	15.58	15.90	16.39	16.10
Fe_2O_3	1.26	1.41	1.12	3.23
FeO	3.22	3.18	2.76	2.18
MgO	3.23	3.33	3.27	3.74
CaO	4.51	4.63	4.07	1.24
Na_2O	4.01	4.05	3.84	4.78
K_2O	2.22	2.10	2.08	1.69
H_2O at 110°	0.19	0.18	0.22	....
H_2O above 110°	1.17	1.16	1.82	1.52
TiO_2	.69	.69	.44	....
P_2O_5	.16	.17	.11	....
MnO	trace	trace	.05	....
SrO	trace	none	trace	....
BaO	.11	.06	.08	....
Li_2O	trace	trace	trace	....
S	trace	trace	trace	....
	100.89	100.23	99.98	100.78

No. 1—U. S. G. S. Mt. Stuart Folio, from southeast slope of Mount Stuart Range, light-colored phase of granodiorite (H. N. Stokes).
No. 2—Mount Stuart Folio, Ingalls creek, darker variety of granodiorite (H. N. Stokes).
No. 3—Mount Stuart Folio, from dike on branch of Icicle creek west of Mount Stuart.
No. 4—Sheep Mountain, acid light-colored granodiorite dike, (M. C. Taylor, University of Washington).

*Mount Stuart Folio, U. S. Geological Survey, pp. 2-3, 1906, Washington, D. C.

SWAUK FORMATION.

AREAL DISTRIBUTION.—The Swauk formation occupies an area of roughly 1,000 square miles in the eastern part of the Cascade* mountains and west of Columbia river. The belt is irregular in shape, but has an average width of twenty miles east and west and approximately fifty miles north and south, and extends from beyond Wenatchee lake on the north, southward across Wenatchee river and from thence up over the Wenatchee mountains down to the Yakima valley. On the west it lies unconformably upon the older metamorphic and igneous rock and on the east it passes beneath the later Tertiary lavas, as in the vicinity of Table mountain. Locally, within the area of the Blewett district, it outcrops as one large mass, having an area of nearly one square mile in the northeastern corner. In the southeastern corner and the southwestern corner there are two small patches, which beyond the limits of the district are areally connected.

GENERAL DESCRIPTION.—The Swauk formation represents a succession of beds of conglomerate, arkose sandstone, and shales, with occasional carbonaceous seams, all of which are more or less interstratified. No broad subdivision of this formation into characteristic upper and lower divisions can be made, except that the massive granite conglomerate generally represents the base. However, the conglomerates which form the base at one locality may correlatively represent some higher member of the series at another locality in the same formation. These beds were deposited in an extensive but irregular shaped freshwater lake during early Eocene time, and owing to the differential warping of the floor of this lake during the process of deposition of the sediments, the character of the materials laid down at the same time varies because of differences in the depth of water in different locations, and the proximity to shore lines. Hence the stratigraphic succession of beds vary from one locality to another, and

---

*Russell, I. C., Preliminary Paper on the Geology of the Cascade Mountains in Northern Washington, 20th Annual Report, U. S. Geological Survey, 1898-1899, Pt. II.

the correlation of the several members in different localities cannot be accurately made.

From studies made in several widely scattered areas, it has been estimated that the Swauk formation aggregates a total thickness of 5,000 feet.*

Because of the changes in the area of deposition, during the time that the sediments were being laid down, the various sections would present different thicknesses and when the fact is taken into consideration that subsequent to deposition there was extensive deformation and that a large part of the formation has been removed by erosion, it becomes apparent that the original thickness is not now represented. Hence stratigraphic sections constructed in different localities represent only portions of the entire stratigraphic column of the Swauk formation.

The base of the formation, wherever observed, rests unconformably upon Pre-Eocene rocks. These basal rocks are composed of materials belonging to the Peshastin, Hawkins, the basic peridotite intrusives, and the granodiorite. They have been extensively metamorphosed and the peridotite had been altered over into serpentine and then eroded, prior to the deformation of the region and the formation of a large irregular drainage basin which developed into an extensive fresh water lake. The basal contact rocks consist of conglomerates ranging in size from fine sandy grits to bowlders which are ten feet or more in diameter, and only partially rounded and water-worn. These variations in character give a fairly clear picture of the conditions of deposition. Many of the large bowlders in the extremely coarse conglomerate consist entirely of granodiorite and indicate local, narrow embayments of the lake basin, where the adjacent land masses were being rapidly torn down and eroded and the materials accumulating close to shore on a rapidly submerging lake floor, so fast in fact, that insufficient time elapsed to allow the action of the water to round the pebbles and bowlders. They now represent a recemented mass of granodiorite bowlders, which are often extremely difficult to distinguish from

*Mount Stuart Geological Folio, p. 5.

A—General view of Meteor and Peshastin tunnel openings in Culver gulch.

B—Twenty stamp mill at mouth of Culver gulch, looking west. Tailings bin in foreground.

the granodiorite mass itself. In other places the basal conglomerates are made up of well rounded water-worn pebbles averaging from one to six inches in diameter, and from that they range down to fine grits. The pebbles comprising this phase consist of materials derived from all of the older rocks immediately underlying, and from those formations which are not far distant. These consist of slates, schists, cherts, quartzites, intrusive dike rocks and the more or less metamorphosed lavas. In some places many of the pebbles consist of serpentine, indicating that the serpentinization of the peridotites and other basic intrusives had already occurred.

In some instances, black carbonaceous shales are interbedded with the sandstones, but none of them appear to be of commercial importance. In one locality near Shaser creek the basal conglomerates and sandstones are cemented with magnetite. This belt ranges from twenty to sixty feet in thickness and consists in some places of serpentine, containing magnetite, with occasional water-worn pebbles scattered through it. From that condition this belt grades into one where the pebbles become more numerous and ultimately form a black appearing conglomerate entirely cemented by black iron oxide, and this conglomerate when traced upwards into the sandstone finally has for its cementing material silica and ultimately grades into an arkose sandstone or shale.

The middle and upper portions of the formation vary in different places, but in a broad general way consist of massive beds of light gray arkose sandstone, interstratified with layers of shale and occasional beds of fine conglomerate and carbonaceous material.

Within the limits of this district the Swauk formation occurs in three distinct localities all of which are areally connected outside the boundaries of the region. These are represented by an area consisting of about one square mile in the northeast corner of the district, and by two very small patches, one in the extreme southeast corner and one in the southwest. In each of these areas the base is represented, or at least the

—4

basal beds for that particular locality. The largest of the outcrops is in the northeastern corner and lies east of Peshastin
creek and constitutes the greater part of the rock in which Ruby
creek has been eroded. The major portion of the Swauk formation as here represented is locally a light gray, coarse, arkose,
massive sandstone showing distinctly the grains of quartz, feldspar, biotite, and muscovite. Very little shale or carbonaceous
matter is interbedded, indicating that the sediments had been well
washed on the beach where they had been exposed to the action
of the waves. The basal members rest unconformably upon the
Peshastin, Hawkins, and peridotite formations. While much of
the material derived from these formations composes the pebbles
and cementing materials of the conglomerates, still the larger
part is made up of pebbles and bowlders of the granodiorite.
About one-half mile to the north of the map, on Peshastin creek,
this conglomerate outcrops on the roadside and represents the
typical, bowlder basal conglomerate previously described. The
contact as observed between the serpentine and the Swauk near
the surface was nearly vertical, or at least with a very steep pitch
to the northeast. The contact plane is irregular, showing little
knobs of serpentine projecting into the sandstone. Local faulting and slipping has also taken place along the contact, and a
good example of this may be seen on the ridge between Negro
and Peshastin creeks. The main mass of the formation is composed of light gray, coarse sandstone, massive in character, with
very few interbedded shales.

In the southeast corner of the map the sandstone is similar
and lies unconformably upon the serpentine. In the southwest
corner it rests upon the eroded edges of both the serpentine and
Hawkins. It is there composed of conglomerates and typical
sandstones. In places the conglomerate is cemented by iron
oxide, giving rise to a band or belt of low grade magnetite
deposits, locally known as "The Iron Dike." A larger portion
of this dike lies just to the south of the boundary of the district
and on the north side of Shaser creek.

CORRELATION AND AGE.—Within the boundaries of the map no fossils were found. However, at many places throughout the region the Swauk does outcrop and fossils have been found, which are represented by leaves that fell from trees growing along the shores of this lake during the period of its existence. Collections of these were made by Dr. George Otis Smith, who submitted them to Dr. F. H. Knowlton. His determinations* are included in Dr. Smith's report and are here included.

The following genera were collected:

Lygodium	Quercus	Diospyros
Sabal	Ficus	Celastrinites
Myrica	Cinnamomum	Zizyphus
Populus	Prunus	Phyllites
Comptonia		

"Of these species a form or variety of only one was previously known, this being *Ficus planicostata*, which is a common species in the Denver and Laramie. Other forms, however, have a more or less close resemblance to certain Laramie, Denver, and Fort Union species, and on this rather insecure basis it is assumed that the age should be regarded as Eocene."

GABBRO.

DISTRIBUTION AND DESCRIPTION.—Numerous intrusive masses of gabbro occur in a belt extending west from Blewett and south of Ingalls creek.†

They do not occur as one continuous mass but rather in several irregular isolated patches, ranging from one to several square miles in area. They have been intruded into the Peshastin, Hawkins, and Swauk formations, and because of their greater ability to withstand weathering as compared with the older rocks, they stand out prominently and form a large part of the ridge south of Ingalls creek and north of Negro and Stafford creeks. Within the Blewett District the gabbro occurs in the northwest part and occupies an area a little less than one-half square mile, where it forms the divide between

*Mount Stuart Geological Folio, p. 5.
†Mount Stuart Geological Folio, p. 6.

Ingalls and Negro creeks.   Here it is intrusive into both the Peshastin slate and serpentine.   Just south of the main mass there are two very small apophyses or dikes intrusive into the Peshastin and connected with the present magma not far below the surface.   One very small outcrop of coarse grained gabbro occurs above the United States Mineral Monument at the head of Culver gulch at an elevation of about 4,400 feet.   These are the only known occurrences of rock which can be definitely assigned to the gabbro intrusions of Eocene time. Many of the partially altered peridotites resemble gabbro and it is possible that they may represent certain phases of the gabbro intrusives.   However, these individual outcrops are so small that they have not been mapped but rather grouped in with the peridotites and serpentine.   They are also partially altered to serpentine.

PETROGRAPHICAL DESCRIPTION.—The gabbros occurring in this region are fairly constant in character.   Most of the hand specimens examined were somewhat altered.   They are medium to fine grained rocks, with a light greenish gray color and to the naked eye appear to be composed of light colored feldspars with green and brown minerals scattered among them.   Under the microscope these white minerals are seen to be largely basic plagioclase whose composition ranges from Ab1 An5 to Ab1 An6.   These crystals average about 1mm. in diameter and sometimes show zonal structure.   They are distinctly allotriomorphic and constitute about one-half the mineral content of this rock. The darker minerals were found to be hornblende, augite and diallage.   One specimen collected from the summit of the ridge between Negro and Ingalls creeks at an elevation of 4,700 feet consists of about 40% diallage and 10% each of augite and hornblende.   Other slides collected further down the ridge near the serpentine contact contained a larger proportion of hornblende, about the same amount of augite and about 10% of diallage.   A small amount of what appears to be olivine is present but partially altered to serpentine.   Small grains of magnetite are very common.   In other slides much of the feldspar is altered to kaolin. Some of the smaller dikes of gabbro cutting up into the Peshastin formation on the north side of Negro creek, have a distinctly ophitic structure and contain hypersthene. The gabbros proper are typically holocrystalline in structure.

DIABASIC DIKES.

AREAL DISTRIBUTION.—Among the prominent geological features represented on the areal map are numerous basic dikes having a general northeast and southwest trend. Their average strike approximates north 25° east. On the eastern side of the Cascades, between Wenatchee river on the north and Yakima river on the south, these dikes outcrop prominently, which may be seen by referring to the areal geologic maps of the Mount Stuart and Snoqualmie folios. Their northeast to southwest strike predominates. These dikes cut through the Peshastin, Hawkins, peridotite, granodiorite, and Swauk formations. Studies made outside of this district give proof that they extend upwards and connect with the lower portion of the Teanaway basalt lava flows, affording ample evidence that those basaltic lavas represent accumulations of basic igneous material which welled up through innumerable northeast southwest fissures and spread over the surface during middle Eocene time. Chemical analyses* made of the diabase dikes and of the Teanaway basalt show them to be genetically associated.

Within the area of this report the Teanaway basalt does not at present exist, but may possibly have done so and have been removed by subsequent erosion. These dikes extend downward nearly vertically and range in width from one foot to 200 feet, varying somewhat with the character of the formation into which they have been intruded. Owing to the original uneven character of the fissures in the serpentine, these dikes vary from point to point, but in the Swauk sandstones and older rocks they are more persistent owing to their harder and more resistant material.

Three prominent dikes outcrop near the head of Culver gulch.

*Mount Stuart Geological Folio, p. 6.

The one on the north side outcrops at an elevation of 3,840 feet and stands up above the surrounding serpentine. This dike is approximately 700 feet long and has an average width of about seventy-five feet. To the south it tapers down to about twenty feet and then pinches out entirely. At the north end it bends and strikes off 60° west of north for a distance of 300 feet and then disappears. The surface rocks on the south slopes of the canyon are covered with large and small angular blocks broken off from the main diabase mass. An interesting structural feature may be observed at the point where No. 5½ tunnel intersects this dike. The tunnel has been driven on the vein and the latter at the point of contact breaks up into numerous stringers of blackened quartz, showing that the veins existed prior to the intrusion of the dike. On the south side of the gulch, at an elevation of 3,400 feet a narrower dike extends in the direction north 70° west for a distance of about 1,000 feet. The north end of this dike bends to the east and pinches out, and apparently cuts the Pole Pick vein near the three tunnels on the fraction. While these two dikes do not rank among the largest in this area, yet they are important because of their structural relation to the ore deposits to be described later.

Several very prominent dikes of diabase occur near the head of Bear creek, one of these forming the western side of Iron mountain, so named from the basic character of the rock. In the ridges on the south side of King creek there are three conspicuous dikes cutting up through the peridotite and granodiorite. The middle one shows well the irregularity of the dike as it passes from the former into the latter. The fissures through which the molten lava came up to the surface were comparatively wide in the peridotite, while in the granodiorite they were much narrower. This is well shown at the contact where the diabase cuts the other two. Short, narrow wedges of the diabase jut into the granodiorite, which is much fractured and broken, and the portion which extends through it is represented by a long narrow band. In places, in the canyon of King creek, it often becomes impossible to trace its continuation except by connecting isolated projections outcropping through the overlying talus material. On the east side of Peshastin creek and north of Sheep mountain three of these dikes may be seen trending north 70° west, and outcropping in the peridotite. On the same side of the creek, on the ridge leading out to Windmill Point a small dike outcrops, and further north a well defined one about a mile in length crosses Ruby creek and pinches out to the southwest near the contact of the Swauk sandstone with the peridotite. One lone but prominent dike traverses the gabbro on the divide between Negro and Ingalls creeks, and trends in a nearly north to south direction. In many places very small, isolated diabasic outcrops occur but the difficulty in tracing them and distinguishing them from the slide material renders it impossible to represent them on the geological map.

PETROGRAPHICAL DESCRIPTION.—Microscopically, this rock upon fresh exposure has a fine grained and very dark gray colored appearance. It has a high specific gravity and breaks with a conchoidal fracture, and rings when struck with the hammer. The weathered surfaces nearly always present a rusty red color, due to an iron oxide stain, and the surface of the hillside in the vicinity of the dike has a reddish brown appearance due to the oxidation of the disintegrated talus material. Upon close examination with the eye very small crystals of augite can be detected and occasionally larger prophyritic crystals of olivine. Often the clean fractured surface is thickly spotted with brown, lustrous specks, showing strong pleochroism, which appear to be the mineral iddingsite.

Upon examination with a hand lens phenocrysts of feldspar can occasionally be detected but they are universally small. Magnetite sometimes may be seen. In the wider dikes the texture is coarser, specially near the middle, but extremely fine grained toward the edges. All of the specimens collected within the area of the map are fine grained, and have a typical diabasic structure.

Twelve thin sections were chosen from the most representative hand specimens and the following minerals were found to be present: magnetite, apatite, augite, orthorhombic pyroxene, olivine, plagioclase feldspar, and occasionally quartz.

Apatite is not a common constituent and is only sparingly found. It occurs both in long slender and short prisms and seems to be most closely associated with the feldspar and augite. It can be detected by its low, double refraction, high index of refraction and lack of color.

Magnetite and possibly ilmenite were found in all of the thin sections and in several of them in great abundance. They appear as minute and opaque grains scattered in irregular masses through the rock without any direct relationship to the other minerals.

Augite, next to feldspar, is the most important constituent of the rock and is generally fresh and unaltered. It has a pale brown color with faint pleochroism. On several crystals which happened to be cut nearly parallel to the clinopinacoid, extinction angles of 49° were obtained. Many of the crystals commonly show a zonary structure. Cleavage is well developed, showing as straight lines in sections parallel 010 and 100 and intersecting at angles of 88° on sections parallel to the base. None of the crystals show distinct boundaries but rather exist as irregular patches with more or less developed crystal outlines.

Olivine does not occur abundantly but where represented shows as clear, fragmental crystals, much decomposed to iron oxide. Occasionally phenocrysts attaining a diameter of 2 mm. were noticed, giving pinacoidal and dome outlines as well as straight extinction.

The feldspar present is exclusively of a plagioclase variety and consists of extremely basic labradorite, occurring mostly in the form of microlite laths. The maximum extinction angles measured as high as 30°. These small crystals form a network and enclosed between them are small grains of augite, giving the typical insertal structure. Quartz, when present, occurs in very minute grains scattered about along with the augite between the feldspar laths. No chemical analyses of this rock were made but one made by the U. S. Geological Survey and published in the Mount Stuart Folio* is here inserted.

ANALYSIS OF TYPICAL DIKE ROCK FROM NEAR NORTH FORK
OF TEANAWAY RIVER.

Compound.	Per cent.
SiO_2	57.21
Al_2O_3	12.99
Fe_2O_3	3.28
FeO	10.18
MgO	1.59
CaO	5.97
Na_2O	3.07
K_2O	1.61
H_2O at 110°	.68
H_2O above 110°	1.03
TiO_2	1.72
P_2O5	.44
MnO	.24
NiO	trace
SrO	trace
BaO	.06
Li_2O	trace
FeS_2	.13
	100.20

QUATERNARY.

ALLUVIUM.—Throughout the eastern portion of the Cascade mountains the Quaternary deposits consist of gravels, sands, silts, and talus deposits, and occasionally large irregular bowlders. These have been eroded away from the high mountain masses, crushed, ground and carried down stream during heavy rainstorms, and finally assorted and deposited over the river flood-plains when the waters receded. Accompanying this process during Quaternary times there was a differential uplift and depression of the land surface with reference to sea level, causing streams to have locally increased or decreased gradients. This produced a silting up of stream valleys in some places and in other places a downward cutting of the stream into the valley floor, allowing the original valley deposits to remain above the level of the stream, forming stream terraces. Nearly all of the larger valleys leading down from the Cascades are mantled over with glacial debris, left upon the retreat of the glacial ice streams. In many cases, moraines may be found showing the

*Mount Stuart Geological Folio, p. 6.

position of the ice as it advanced and finally retreated. Hence all these Quaternary deposits may be classed under three main subdivisions: Terrace gravels, glacial deposits, and valley alluvium.

Within the area of this district glacial deposits are absent. Long tongues of ice came down Ingalls creek, passed into and flowed down Peshastin creek, as may be seen in the moraines along the valley, and in the presence of numerous large granite bowlders, sometimes striated and often perched in precarious positions on other rocks. Above the junction of Ingalls and Peshastin creeks, on the latter stream, no evidence of the action of glaciers could be found, hence it is presumed that the glacial ice never occupied the area involved in this discussion.

Terrace gravels do occur, but not at any considerable elevation above the present valley floor. They may be seen on Peshastin creek below Blewett, extending down as far as Negro creek to the north end of the district. In many places the gravels are mixed with talus materials composed of angular fragments of slates and cherts, making it often impossible to distinguish between the boundaries of the two. On Peshastin creek the pebbles forming these gravels range from the size of a pea to over six feet in diameter, and are composed of materials derived from all of the rocks involved in the drainage basin of Peshastin creek and its tributaries above the place of deposition. Many of these gravels contain placer gold, which will be discussed later in the chapter on ore deposits.

The stream gravels occur along Negro creek to the extreme western end of the district. In several places the width of the valley gravels is several hundred feet. In the western part, on Negro creek, they rest directly upon the Hawkins formation, and east of that upon the Peshastin slates. This is true on Peshastin creek below Blewett and at the junction of the two creeks where Negro creek is cutting a canyon-like gorge in the slates, preventing the local accumulation of gravel until the gradient decreases and the cutting becomes less rapid. Above Blewett the gravels rest upon the serpentine and granodiorite, and about one-half mile south of the map are confined entirely to a bedrock composed of the various members of the Swauk formation. As a rule the gravel lying directly upon the bedrock is fine grained and most of the larger bowlders are found near the surface. All of the smaller creeks emptying into Peshastin and Negro creeks contain more or less detrital material, partially formed by the action of running water, and partially from material collecting on the talus slopes.

STRUCTURE.

The most prominent structural features developed within the Blewett district have been produced by the intrusion of igneous masses and by the deformational movements which this region has undergone along with other portions of the Cascades. The most important structural results of igneous intrusion are the isolation of huge, irregular-shaped blocks of the Hawkins and Peshastin formations, disrupted from larger masses at the time of the peridotite intrusions.

FOLDING.—To a certain extent the old metamorphic rocks have been thrown into a series of anticlines and synclines. While the evidence is not direct, yet it points to the fact that the Peshastin and Hawkins formations are involved in a shallow syncline in the southeastern portion of the map. In the vicinity of Sheep mountain, the Hawkins breccias and tuffs pitch down to the northeast while directly below and south the Peshastin slates are doing the same at the low angle of about 15°. Just north of Windmill point the Peshastin is pitching to the south at a very steep angle. Immediately south and above this locality the Hawkins formation outcrops. The contact is separated, however, by an irregular band of partially serpentinized peridotite. If such a structural feature does occur, large portions of the syncline, including the axis, have been torn asunder by the intrusion of the peridotite. On the north side of Negro creek the Peshastin formation apparently pitches southward beneath the Hawkins. The Swauk sandstone in the southeastern and southwestern portions of the district dips away from the other formations at a very steep angle, and in several places only a short distance farther south it stands nearly vertical.

FAULTING.—No pronounced faulting has been discovered. Many small faults exist which do not seem to be of sufficient extent to be distinguished on the map. They occur mostly in the serpentine. The fissure veins extending east and west along Culver gulch represent faulting and fissuring which occurred early in the geological history of the region, but which were subsequently filled by mineral bearing veins. Suggestions have been made that a later east-west fault occurs near the head of Culver gulch, parallel to the most prominent veins, and which has resulted in the dislocating of two diabasic dikes, one on the north side of the gulch from that on the south. A detailed examination, however, seems to indicate that these were not a part of the same dike, but rather two individual dikes which have come up through two separate openings. The end of the south dike tapers off to a point and the end of the upper dike on the north side seems to narrow down and pinch out. One small, nearly horizontal fault was observed in the discovery tunnel in the patented Pole Pick No. 1 mining claim. The quartz vein was dislocated about seven feet and the plane of faulting appears to be nearly horizontal. This fault probably represents merely a local break or slip and is not important. In nearly all of the tunnels passing through the serpentine formations, so-called walls or slips are frequent, which pitch and strike in various directions. The larger number of these, however, were apparently produced at the time of the serpentinization of the peridotite. Many of the fissure veins show secondary movement since their deposition, as may be seen in their grooved and slicken-sided walls.

GEOLOGICAL HISTORY.

The geological history of the Blewett Mining District is a part of the geological history of the Cascade mountains. The same diastrophic movements involved in the deformation of the Cascades during the various periods of its history, the changes in the relative positions of land and sea, the metamorphism produced by igneous intrusions, the deposition of sedimentary rocks in the large fresh water lake basins, and the final modification of the topographic features by the action of stream erosion, which has operated upon so tremendous a scale in developing the Cascades as a whole, have been the factors involved in producing the geological and topographical features in the area under discussion.

The formations in this region may be divided into two series which are separated by a pronounced unconformity. The older of these is represented by a bedrock complex, consisting of metamorphic and igneous rocks of Pre-Tertiary age. The younger of the two series is composed of non-metamorphosed sedimentary rocks together with igneous intrusive and extrusive rocks. The earliest records of the geologic history of this district must be sought for in the rocks of the Peshastin formation. The geological period during which this formation was deposited cannot be definitely determined by the evidence obtained within the area of this map, but it corresponds most closely to formations occuring in British Columbia, known as the Cache creek series, and to the Calaveras formation in the Sierra Nevada of California. In other portions of the Cascades, rocks similar to these occur. In British Columbia and California the formations most closely related to these are known to be Carboniferous, and based upon that evidence, the Peshastin formation provisionally may be assigned to the Carboniferous or older.

During Peshastin times, from all of the available evidence obtainable, this region was a part of the ocean or at least an arm of the ocean, and in it were accumulating conglomerates, sandstones, shales, and limestones. The pebbles forming the conglomerates were derived from the erosion of adjacent land masses, composed of earlier metamorphic rocks. In the deeper water far from the shore the shales and limestones were being laid down. During the time these were being deposited there appears to have been more or less crustal deformation, causing a part of the region to be slowly elevated and a part depressed. After a time volcanic activity became the important factor in the history of this region, and molten basic igneous rocks found their way to the surface, possibly from volcanic centers situated on some adjacent land area, or possible located upon islands which may have existed in the sea. At any rate, accumulations of volcanic ash and scoriae were deposited in the open waters, and then worked over by the waves, and intermixed with fragments of chert, greenstone, and slate, derived from the land masses along the shore. These were followed by outpourings of volcanic lava which picked up fragments of rocks composing the floor over which they were flowing, and carried them along in the current, giving rise to an aggregate of volcanic breccias. The precise order in which these disturbances took place cannot definitely be determined, but the conditions just described characterize this period of time and are the interpretation of facts gathered in the field. These materials constitute what are known as the Hawkins formation. Within this formation are irregular masses of diabase and gabbro porphyry, which indicate that a part of the lava at least may have reached the surface through fissures, and which upon cooling resulted in the formation of basic dikes. The two formations just described constitute a part of the Pre-Tertiary metamorphic complex and are separated by an unconformity, but neither the exact conditions nor the length of time intervening between them could be determined.

After the deposition of these two formations, deformational movements became a prominent feature in the history of the region, resulting in the uplift of the recently accumulated materials into the form of a mountain mass. The geographic conditions existing at this time cannot be definitely known because of the comparatively small area of these rocks exposed at the present surface. It seems probable that the region was undergoing erosion for a long time, and that the sediments were being carried off and deposited in some basin which is now deeply buried by later rocks. The next epoch was inaugurated by the intrusion of great plutonic masses of ultra-basic rocks. They were largely of the nature of peridotites and basic gabbros, the latter apparently being a differentiation product from the underlying magma. The time involved in the intrusion of these basic rocks must be conceived of as very long, during which there was not one single intrusive mass injected up into the earth's crust but rather a series of differentiations from a composite, basic magma deep below. We may consider first that those portions of the magma which were richest in iron and magnesia were drawn off from the underlying reservoir and injected into the earth's crust, and then to have solidified, forming a rock composed essentially of olivine and enstatite. Later more acid phases, richer in calcium and silica were drawn off from this same underlying magma and forced up into the already solidifying rock. The later intrusions were in the nature of basic gabbros and appear in the field as irregular shaped dikes, whose boundaries vary and gradually grade into peridotite. The age of this series of basic intrusions cannot definitely be stated. It is a known fact, however, that intrusions of peridotite were a characteristic feature of the Upper Jurassic near its close. Throughout the Sierra Nevada and the coast ranges of California* and Oregon they are known definitely to be Pre-Tertiary, hence the age of these intrusions may be provisionally assigned to the Jurassic. During Cretaceous time we have no evidence of sedimentation, and apparently none of volcanic activity, unless it be intrusions of granodiorite, which, however, may be of Jurassic age.

Subsequent to the time of the peridotite intrusion and prior to the deposition of the Eocene sandstones, this entire region was invaded from below by a great batholith of granodiorite, which is now represented by extensive outcrops in the vicinity of Mount Stuart and northward, and by numerous smaller dike-like apophyses cutting up into the peridotite and older metamorphic rocks. Accompanying these intrusions, there was much deformation and fissuring of the overlying rocks. From this slowly cooling, deepseated granodiorite magma, solutions under extremely high pressures and temperatures were gradually circulating upwards through the peridotite, and extracting calcium and iron

---

*H. W. Turner, Geology of the Sierra Nevada; 17th Ann. Rept. U. S. Geological Survey, Pt. I, p. 549, 1896.

from the calcium-ferromagnesium silicates, and in turn yielding up a part of its water to unite with the magnesium and the silica to form the hydrous magnesium silicate serpentine. This chemical alteration of the peridotite increased the volume of each individual crystal, causing much local crushing and shearing. It seems evident that these same solutions which were causing serpentinization also carried silica, gold, arsenic, iron, and sulphur, directly from the slowly solidifying granodiorite magma to the downward extending east-west fissures trending through what is now Culver gulch. Upon the release of pressure and the decrease in temperature, the mineral contents along with the silica and calcium carbonate were precipitated from the solution, forming the veins as we know them today in the deepest workings in Culver gulch. After these ore bodies were formed the region underwent extensive deformation and erosion until the beginning of the Eocene.

The Tertiary history of this region was inaugurated by intense deformation of the surface rocks, resulting in the formation of a great lake basin. The first material to accumulate in the form of deposits upon the floor of this lake consisted of a basal conglomerate, made up in some places of fine pebbles of serpentine, slate, and greenstone, and in other places of coarse bowlders of granodiorite. Succeeding these, as the lake enlarged and deepened, were sandstone, shale, and occasional carbonaceous beds. In the sands and muds were buried leaves fallen from trees, growing along the shores. These sediments are known as the Swauk formation. Following the deposition of this formation, volcanic activity again became an important geological factor, resulting in the intrusion of basic magmas through the numerous conduits leading up to the surface, and then spreading out as great basalt flows. The flows which are known as the Teanaway basalt do not occur within the limits of the area mapped, but are well represented only a short distance to the southwest. A little before the intrusion of the diabase dikes masses of gabbro were injected into the earth's crust, and apparently are genetically related to the dikes. After the outpouring of this material had ceased, this portion of the Cascades again underwent extensive deformation, resulting in the formation of a second large lake, in which accumulated the Roslyn formation of Upper Eocene age, but which did not extend as far northward as the Blewett district. During the remainder of the Miocene and Pliocene times extensive flows of lava were accumulating in the Cascade mountains, not far distant from Blewett, but as far as we have been able to determine, never covered this immediate region. During this long period of erosion, however, the region was reduced to a peneplain, and then at the close of the Pliocene or beginning of the Quaternary, the Cascades were again uplifted to at least 8,000 feet above sea level, by a series of complex deformational movements. The most prominent of these movements, and the one primarily affecting this district was the development of a long axial uplift or warp in the earth's crust trending southeast to northwest from Columbia river to Mount Stuart and known as the Wenatchee mountain uplift.

The physiographic processes involved during the Quaternary have already been described in the chapter dealing with the topographic development of the region. Briefly, they represent a gradual uplift of the Wenatchee mountain mass followed by the development of new drainage lines, the cutting of canyons by streams, the adjustment of the streams to differences in the character of rocks encountered and finally the modification of this newly developed topography by the advance and retreat of extensive glaciers. The very last events in this history are the geological and physiographic processes that are going on at the present time. Among these may be mentioned the accumulation in the valleys of the stream gravels, and alluvium, the former sometimes being gold bearing.

—-5

CHAPTER III.

ECONOMIC GEOLOGY.

HISTORY OF MINING.

Little is known of the earliest history of the Blewett Mining District. The first mining records are those of several prospectors who, upon returning from the Caribou and Frazer river districts, of British Columbia, and the Similkameen in northern Washington, wandered southward into the eastern foothills of the Cascades and while searching on Peshastin creek, in 1860, discovered rich placers. After working for a time the first prospectors migrated southward and the work was continued by those following them from the north. Very little information is at hand concerning these early pioneers. One of them, a negro, is said to have worked the gravels on Peshastin creek near the mouth of Negro creek, named for him, and to have taken out about $1,100. The discovery of the first quartz vein in this district is said to have been made by one of the United States soldiers in 1854, but it was not located.

About 1874 the first quartz claim was located by John Shafer near the head of Culver gulch and has since been known as the Culver claim. About the same time the Pole Pick, situated on the south side of Culver gulch, and the Hummingbird, were located by Samuel Culver. Shortly afterwards James Lockwood located the Bobtail, John Olden and Peter Wilder the Fraction, and John Olden and Samuel Culver the Little Culver.

While these locations were being made in Culver gulch, what was considered to be the same lead was found to extend eastward and upon it were located the Peshastin and Blackjack claims. Still farther east, on the east side of Peshastin creek, were located the Golden Chariot and Tiptop claims. To the west of Culver gulch, trending westward down to Negro creek, were located the Shafer, Vancouver and Seattle claims, by Marshall Blinn. Later these were sold to the Cascade Mining Company. In addition to these, many other locations were made on the Pole Pick and the North Star ledges, and others both to the north and to the south of the east-west mineralized zone.

In the very early days, Blewett was connected with the outside world by trail only, but in 1879 a wagon road was built from Cle Elum on the south, over the Wenatchee divide to Blewett. The ore bodies first opened up in the development of quartz mining were mainly those from the oxidized zone, which were free milling. Several arrastras were built for treating these ores, two of which have been in use until very recent years. Late in the seventies a six-stamp mill with one Frue vanner was erected and operated by water power. For about eight years this mill was run and the ores treated in it were obtained from the upper tunnels on Culver gulch. In 1891 a ten-stamp mill was erected by the Culver Mining Company, who sold it the following year to the Blewett Mining Company, who again resold it in 1896 to Thomas Johnson. In this mill were treated a large part of the ores from the Pole Pick, Tip Top, Blackjack, Peshastin, and other claims lying on the ridge between Culver and Culver Springs gulches. Shortly before the sale of this ten-stamp mill to Thomas Johnson, the Blewett Company erected the present twenty-stamp mill at the mouth of Culver gulch and began a systematic underground development of the property. From 1894 to 1897 the practice was made of leasing certain portions of the mines on a royalty basis, the result of which was the opening up and stoping out of the rich oxidized ores in the upper portion of the gulch. In 1896 the Warrior General Company was organized, which paid $35,000 to the Blewett Company for its property. Shortly afterward this company was reorganized and called the Chelan Mining and Milling Company. About this time extensive development of the Peshastin ore bodies was begun by driving a series of south cross-cuts from Culver gulch to the vein. Rich bodies of ore were encountered and for a time the Blewett district became very active. In 1905 this property, together with the La Rica, became the Washington

Meteor Company. During the last ten years the greater part of the work has consisted in the development and mining of the ores from the Meteor and Peshastin tunnels on the Peshastin claim and from No. 9 tunnel on the Culver claim. In recent years several properties have been patented and those which have not have confined themselves mainly to assessment work.

The placers in this district have been worked from time to time by individuals and in 1909 extensive plans were made to mine the gravel on Peshastin near the mouth of Negro creek by sluicing with water brought by flume from further up Negro creek. Because of poor management this did not prove a success and was abandoned.

During the summer of 1910 sixteen claims were located on a series of iron oxide outcroppings situated to the south of the Blewett district. Since these claims are not a part of this report, their description will not be undertaken.

TREATMENT OF THE ORES.

Up to the present time the ores of this camp have been considered as free milling. The greater part of these have been mined from the upper oxidized zone and consequently have been treated by crushing and amalgamation. The earliest method of treating these ores was by means of arrastras which were used intermittently up to within a very few years ago.

Two arrastras, situated near the mouth of Culver Springs gulch, are still in existence but are unused. Remnants of older ones may be seen on Peshastin creek and in Culver gulch. The best preserved one is owned by Mr. John Olden and is represented on Plate No. VB. The pit is twelve feet in diameter, three feet deep, and built of granite brought from outcrops not far away. The power is obtained by a large overshot water wheel, twenty-six feet in diameter, and run by a stream of water brought from the gulch by a flume and carried directly over the upper portion of the wheel. This operates a horizontal wheel to which granite bowlders weighing over one-half a ton are attached by chains. These are known as drag blocks and crush the ore to a fine powder. The arrastra is said to have crushed from one to two tons of ore per day.

The first stamp mill to operate in the state was erected on Peshastin creek and consisted of six stamps with one Frue vanner, and is said to have reduced eight tons of ore in twenty-four hours, and to have yielded $21 per ton from Culver ore in the first nine days run. After running for eight years this mill was shut down.

In 1880 a two-stamp Huntington mill was brought in from The Dalles, on the Columbia, and erected on Negro creek about two miles above its mouth. This mill was run for about two years and then stopped, never to resume. About fifty tons of base ore, assaying from $10 to $70 per ton, are said to have been reduced, but on account of the large amount of arsenopyrite, pyrite, and iron-copper sulphides it was impossible to save more than $4.50 per ton.

In 1891 the Culver Gold Mining Company erected a ten-stamp mill with four Woodbury concentrators on Peshastin creek near the mouth of Culver gulch and brought the ore from the head of that gulch to it by a cable tram. The following year this mill was sold to the Blewett Gold Mining Company and in 1896 was again sold to Thomas Johnson, who began milling the Pole Pick ore.

In 1892 the Blewett Company erected a twenty-stamp mill at the mouth of Culver gulch, allowing space for twenty more stamps. The equipment of this mill has been described by Professor Milnor Roberts in a previous report on the reduction plants in Washington* and will be inserted here. "The batteries consist of four sets of five stamps each, with Fraser and Chalmers automatic feeders, and double discharge, only single being used. 950-pound stamps, with chrome steel shoes and dies, drop $6\frac{1}{2}$ inches at the rate of 90 per minute. The pulp is screened through diagonal slot screens, equivalent to 50 mesh, and falls on copper plates four feet wide and ten feet long, sloping $1\frac{1}{2}$ inches per foot. The lower plates are silvered, 14 feet long,

---

*1st Ann. Rep. Washington Geological Survey, p. 150, 1901.

4 feet wide, and falling two inches per foot. Four Union tables receive the pulp, after which the slimes pass over canvas tables with a three-inch fall. The canvas is swept four times in twenty-four hours, and the fines are saved in settling boxes. Under former management, the tailings carried values of several dollars, which ran into the creek. Outside parties becoming aware of this, built a small cyanide plant with two tanks, having a capacity of about ten tons per day, and thereby recovered a considerable amount of fine gold. The plant is no longer in use.

Wood is burned under two boilers (4 by 12 feet, used alternately), which furnish steam for a Corliss engine of 50 h. p. A flume 500 feet long brings water from the creek to a tank set 20 feet above the level of the stamp battery. A Hallidie aerial tramway with buckets holding 250 pounds each, carries the ore from the mine 4,000 feet distant and dumps it into two receiving bins of 400 tons capacity. The usual system prevails in regard to the different floors of the mill, the order here being crusher, feeding bin, battery, and concentrating floors." During the year of 1896 this mill is said to have reduced 2,469 tons of ore from the Culver claim, which assayed $12.62 a ton, and in addition to this 473 tons of customs ore; the total product of ore in bullion for the company in the year 1896 being $60,000.

In the summer of 1896, a small cyanide plant with two tanks and having a capacity of about ten tons per day, was erected near the twenty-stamp mill, for the purpose of treating the tailings. It had been found that the arsenic in the ore prevented the quicksilver on the plates from catching the gold and consequently dams were built to catch the slimes. 600 tons of tailings, assaying from $3.00 to $30.00 a ton in gold were accumulated, from which 70 to 75% of the gold was extracted by the cyanide process. Later this was closed until the summer of 1910, when the tailings were again run through by the Blewett Mining and Leasing Company.

During the summer of 1910 assays were taken on the iron stained serpentine rock on the south side of Culver gulch, just above the mill, and values reported sufficiently high to allow profitable mining. Chemical tests were made upon these ores with the result that it was deemed best to treat them directly by the cyanide process. Consequently the mill was dismantled, the concentrating tables taken out and three cyanide tanks of thirty-five tons capacity installed. At the present time it is impossible to determine what success was encountered.

In the summer of 1907 a six-stamp mill, with amalgamating plates, was erected in Culver gulch about one-third mile west of Blewett, by the Golden Eagle Mining and Milling Company. This mill is run by a 25 h. p. Fairbanks-Morse engine. Altogether about 100 tons of ore from the North Star and Golden Eagle claims have been reduced, yielding returns of $5 per ton. The mill was run during the month of July in 1910.

PRODUCTION.

Statistics of the total output of metals from this district are lacking. Gold and silver both occur, but the gold far exceeds in importance that of any other metal. The total production, including both placer and quartz, from 1870 to 1901, has been estimated at $1,500,000 and the greater part of this came from the mines in Culver gulch. From 1901 to 1910 it was estimated that these mines produced about $200,000 in bullion, making the total production of gold to the year 1910 about $1,700,000. According to General J. D. McIntire, who had charge of the management of the La Rica Mining Company's property at Blewett from 1900 to 1905, the ore values varied much but ranged from $3 to $10 per ton, and occasionally rich ore shoots averaged as high as $10,000 per ton.

DISTRIBUTION OF THE ORE BODIES.

The larger proportion of the ore bodies included within the Blewett Mining District occupy a belt approximately three miles in length and one mile in width, extending in a general east-west direction. This belt begins on the east side of Peshastin creek, on the slopes of Windmill point, crosses that creek at the town of Blewett and from there extends up Culver gulch, across the divide and down into Negro creek. Small isolated veins occur throughout the entire district but those which have proven commercially productive up to the present time are confined to the belt just mentioned. The richest ore bodies within this belt have been found in Culver gulch.

CHARACTER OF THE ORE BODIES.

The ore bodies occur in a series of irregular shaped but well defined fissure veins, situated chiefly in the serpentine rock. The vein material consists of quartz, calcite, and talc, carrying gold, arsenic, iron, sulphur, and a very small amount of silver, lead, and copper. In the upper or oxidized zone the gold occurs in the free state in the rusty, crumbly, decomposed iron-stained quartz, often containing much talc, especially along the hanging and foot walls. Occasionally great masses of unaltered quartz are intermixed but generally show numerous fracture zones filled with a coating of yellow iron oxide. In the deeper workings the vein material is more characteristically composed of calcite with quartz heavily impregnated with arsenopyrite and pyrite, carrying gold that ranges in value from one dollar to several thousand dollars per ton. In some of the richer pockets in the lower workings, large flakes and wire-like pieces of gold may be clearly seen with the naked eye. Many of the richest specimens occur in talc along the wall rock.

STRIKE. The predominant strike of the quartz veins in Culver gulch is approximately north 75° west. About half way between Culver gulch and Peshastin creek the entire series of veins swings more decidedly to the southeast.

PITCH. Three parallel veins occupy the fissures in Culver gulch. These are known as the Peshastin lead, the Pole Pick lead, and the North Star or Phoenix lead. The Peshastin, the most northerly of the three, and the one on which the largest amount of development work has been done, pitches predominately to the south at an angle of about 75°. In many places it is nearly vertical, but observations taken along the lateral extent of the vein, from the highest to the lowest workings, show that the southerly dip prevails. The Pole Pick or middle of these veins, near the surface at least, pitches to the north at an angle of about 70° but from observations made in some of the lowest workings there is reason to believe that the prevailing dip is nearly vertical and that the northerly dip near the surface is a part of the irregular curving in the downward extending fissure. The North Star or southern of these leads has been opened at only a few points and those are near the surface, hence no general deduction can be drawn as to the probable direction of the downward pitch. Small veins of no lateral or downward extent which have been worked in other parts of the district in general have an east to west strike but represent no prevailing series of mineral filled fissures.

SHAPE. The veins vary in width both along their lateral extent and in their downward pitch. Their shape and size is dependent mainly upon the original character of the fissure prior to mineralization. It is assumed that three predominant fissure openings were formed and that later sufficient pressure was exerted upon them to cause the walls in some places to meet, thus resulting in a plane of fissuring containing a large number of lens-like cavities which later were filled by mineral solutions. These irregular-shaped lens-like pockets are known as ore shoots and have been the source of the richer ores mined in this district. They vary in width from three to sixteen feet, in horizontal extent from 100 to 600 feet, and in vertical range from 20 to 200 feet. The vein material between these ore shoots generally consists of quartz seams or a mixture of quartz and calcite, carrying low grade ores and ranging in width from two or three feet to two or three inches. Very often these veins pinch out entirely, simply leaving the walls in contact with each other. Occasionally these barren zones are represented by a network of disconnected stringers of quartz and calcite with no well defined wall. Examples of these occur in the Meteor and Peshastin tunnels, in No. 9 tunnel, and in the various openings extending up to the head of Culver gulch, where the big summit pocket occurs. The ores in the summit pocket were mainly oxidized quartz and owe their richness to downward leachment. These old workings are inaccessible, but are said to have pinched out 200 feet below the surface. From their general character they may be regarded as the lower portion of an ore shoot, the upper portion of which had been removed by erosion earlier in Quaternary time. During all of Tertiary time, while this region had been undergoing extensive erosion, it seems probable that several thousand feet of overlying rock had been removed and that the upward extension of the vein originally existed in this upper belt. The deepest workings at the present time are on the level of Peshastin creek and while no definite data have been obtained by drilling, there is every reason to believe that the ore bodies will maintain the same general character and mode of appearance that they do in the lower workings in Culver gulch. No free milling oxidized ore may be expected, but arsenopyrite, and pyrite, along with quartz, calcite, and talc may be expected to carry the gold deposits.

Faulting does not play an important part in the displacement of the ore bodies. In many places the veins show the result of local movement along the zone of fissuring. These have occurred since the deposition of the ore, and are represented by smooth slicken-sided surfaces, more or less grooved and parallel to the direction of movement. Where this occurs the wall rock and vein have been ground up, resulting in a clay-like gouge material which is often filled with talc. These zones commonly represent channels for downward percolating water.

INFLUENCES OF COUNTRY ROCK ON THE ORES.

The ore bodies are found in the serpentine and to a lesser extent in the slates and quartzites of the Peshastin formation, and in the volcanic breccias and tuffs of the Hawkins formation. In the rocks belonging to the latter two formations the veins are more regular in character, but when they pass over into the serpentine they conform pretty closely to the conditions just described. Considering the main line of fissuring, extending through Culver gulch, the westernmost portions of the vein lie in the Hawkins formation and when traced easterly towards the serpentine contact, near the head of the gulch, they appear to widen out and split into parallel connecting fissures. When traced easterly beyond Peshastin creek, they again pass into the

Hawkins and may be seen in the workings of the Tip Top mine, up towards Windmill point.

The character of the ore varies with the country rock. Where the veins occur in the Hawkins breccia there is a larger percentage of quartz forming the gangue material. In the serpentine, calcite plays a more important part. No very deep workings have been developed upon the vein in the Hawkins formation and consequently our information in regard to the character of the ore in depth is meagre. On the lower Blinn tunnel on Negro creek the ore is said to have been quartz with some calcite, containing arsenopyrite, pyrite, and chalcopyrite. The property was inaccessible at the time and only a few specimens of the ore said to have been taken from this tunnel were seen.

ALTERATION OF COUNTRY ROCK.

Along the general trend of the veins in this district the country rock has been considerably altered. On the surface it appears as a long, reddish-yellow belt, due to the staining of the wall rock by iron oxide. Below the surface this altered condition is not so prominent and the walls are found to be composed of serpentine, more or less impregnated with iron sulphides and seams of calcite. These belts vary in width from 10 feet to over 100 feet and are locally known by the miners as porphyry dikes. In many places they are porphyritic, but they represent in such cases the less serpentinized phases of the peridotite. A chemical analysis has been made of a specimen from one of the most typical outcrops forming the north wall of the North Star vein, on the south side of Culver gulch, at the North Star mine tunnel, and is inserted below.

MINERALOGY.

In this district the ores may be divided into free milling, oxidized quartz veins near the surface, and quartz-calcite veins carrying arsenopyrite and pyrite, with gold at depth. The vein material in the oxidized zone consists of white, barren, compact quartz, more or less fractured, broken, and stained with iron oxide, and often showing small flakes of gold. In the lower workings the gangue material is commonly composed of quartz or calcite, banded with chloritic material, and with talc occurring on the vein walls. The gangue is thoroughly impregnated with pyrite, arsenopyrite, and occasional crystals of galena. In the talc free gold is common. On Negro creek chalcopyrite also occurs in the veins whose walls are made up of the breccias of the Hawkins formation. When serpentine forms the walls, as it does in Culver gulch, it is usually heavily impregnated with pyrite and yields a small amount of gold. A chemical analysis of this rock was made from a specimen collected from the north wall of the North Star vein, at the face of the upper North Star tunnel, and is here inserted. (Analysis made by M. C. Taylor.)

No. 1107.

SiO_2	31.88%
Al_2O_3	trace
Fe_2O_3	5.81
FeO	6.21
MgO	26.42
CaO	6.84
Na_2O } K_2O }	0.83
H_2O] above [110°]	0.32
CO_2	22.02
Total	100.33

QUARTZ. Quartz is the most prominent vein material, and generally occurs in a milky white form with occasional bands of pyrite and arsenopyrite running through it, or disseminated through it in small isolated crystals. Free gold occasionally may be seen in small flakes within it. In the oxidized zone it is generally heavily stained with iron and sometimes breaks down into a soft crumbly mass.

CALCITE. Next to quartz calcite is the most prominent gangue mineral. It occurs in greater abundance in the lower workings than in the upper, and has impregnated through it pyrite, arsenopyrite, and free gold.

LIMONITE. This mineral occurs in the upper workings as a coating on quartz and serpentine, and represents the oxidized iron sulphide.

MAGNETITE. This mineral, scattered through the peridotite and gabbro masses, occurs along the contact between the peridotite and Swauk sandstone just south of the district and north of Shaser creek as the cementing material of a basal Swauk conglomerate, and has been derived from the peridotite mass.

COPPER. Native copper is very rare in this district. One small hand specimen taken from a diabase dike at the head of Culver gulch showed a few small particles clinging to it. It is said to have been obtained in small amounts in the lower workings on the Blinn property.

PYRITE. This mineral is the most common sulphide. It occurs sparsely disseminated through the Peshastin slates, through the breccias of the Hawkins formation, and in the serpentine especially near the mineral veins. In the veins proper it occurs as crystals, as well as in the massive form. Occasionally it forms bands in the quartz and calcite. The talc on the hanging and foot walls is heavily impregnated with cubes of pyrite ranging in size from six inches in diameter to those which are microscopic.

ARSENOPYRITE. Next to pyrite, arsenopyrite is the most common sulphide, especially in the lower workings. It occurs in small crystals as well as massive and is directly associated with pyrite and with stibnite. It is generally associated with the ores in the Meteor and Peshastin tunnels, where some of the richer gold ores have been extracted.

CHALCOPYRITE. Chalcopyrite is sometimes found, along with pyrite and arsenopyrite, in the veins on the Blinn property. It is generally associated, when present in any quantity, with the gabbro masses.

GALENA. This mineral is not commonly found. It occasionally occurs in very small cubes, along with arsenopyrite and pyrite, in the veins in Culver gulch.

STIBNITE. This mineral has been found, along with arsenopyrite, in the workings in the Peshastin tunnel but is not characteristic of the district.

MALACHITE. Malachite occurs in the oxidized zone in the gabbro area, in a small shaft on the ridge leading westward from the United States mineral monument at the head of Culver Springs gulch. It occurs as a coating extending through the gabbro mass but has not been found in sufficient quantities to pay.

CINNABAR. Mercury sulphide is found at intervals in the serpentine rocks west of the Blewett area, but within this district has been reported only in the small saddle at the head of King creek. A very narrow seam, less than .1 inch in width was found cutting through the serpentine in an east-west direction, but could not be traced over 20 feet.

GOLD. Gold occurs native in both the oxidized zone and in the lowest workings yet encountered. As a rule it is very fine, so that it cannot be detected with the unaided eye, yet may be seen by pulverizing the ore and panning it. Sometimes large flakes may be seen extending through the quartz, calcite, and talc gangue matter. Some exceptionally rich specimens of talc containing stringers and films of native gold were taken from the rich ore shoots on the Peshastin claim. The placer gold is generally fine. The largest nuggets are said to have had a value of about $9 each.

GENESIS OF THE ORES.

All of the veins in this district are confined exclusively to the Pre-Tertiary formations. They occur mainly in the peridotite and serpentine rocks but extend also into the Peshastin and Hawkins formations. No evidences of ore deposition were found in the lower Eocene sandstones. During the period when the lower Peshastin and Hawkins formations were being deposited the region was undergoing much deformation and finally was elevated above the sea level. Later came a series of intrusions of basic igneous magmas in the nature of peridotite. These intrusions caught great blocks of the overlying rock and engulfed them into its mass. Then followed a period of erosion and later came intrusions from a great underlying batholith of granodiorite, a large part of which is now exposed at the surface in the region about Mount Stuart. Extending up through the serpentine and the other metamorphic rocks are long dike-like apophyses of this same granodiorite. These facts indicate that not very far below the present surface these dikes coalesce and become a part of the underlying batholith. At some time after the intrusion of the peridotite and prior to the deposition of the Swauk sandstones, these peridotite masses were altered to serpentine by processes which were hydrothermal in nature. This process of alteration involves the hydration of the iron magnesium silicates which are the component minerals of the peridotite and thereby causing an increase in volume of the original rock. This increase in volume exerted differential pressure throughout the mass causing many zones of shearing and minor faulting, the result of which was, the development of irregular curved and slicken-sided surfaces. While there is no direct evidence that the granodiorites have assisted in the serpentinization of the peridotites, yet it may be inferred that the heated waters which were gradually being driven off from the slowly cooling underlying magma were responsible.

It has been suggested that the ore bodies may have been derived from the solutions genetically associated with the intrusions of the Eocene diabasic dikes, because of the direct origin of the ores in the Swauk mining district from these dikes. The observations made at the contact between the diabase dikes and the Culver vein in Culver gulch indicate that the dike is younger than the vein and, if this be true, the period of ore deposition in this district must have occurred at the time or just after the intrusion of the granodiorite and prior to the intrusion of the diabase dikes. If this theory is correct there is a suggestion that the same heated solutions which were derived from the granodiorite and yielded the water necessary for the serpentization of the peridotite may have also acted as the mineral bearing solutions, which upon reaching the more prominent fractures and fissures deposited the ores. It would seem very probable that much extensive faulting would occur accompanying the deformation produced by the intrusion of the granodiorite magma and that these faults would extend downwards perhaps even to the magma itself. Considering that at this same time the peridotite was undergoing serpentinization, one would expect that any open fissures which might exist would become partially closed due to the pressure exerted by the hydration and increase in volume of the peridotite resulting in a fissure with the walls in contact but leaving lens-like openings more or less connected throughout the original plane of faulting. Under these conditions the heated solutions emanating from the granodiorite magma below and heavily charged with their mineral content would, after slowly permeating through the peridotite mass, find their way to these openings and then slowly circulate upwards toward the surface and under reduced pressures and a lowering of the temperature, the mineral content along with the silica would tend to crystallize out. The first precipitation would occur as an incrustation along the walls of the cavities and the fissures and as more and more material was brought by the upward circulating solutions, continued precipitation would ultimately fill all of the open spaces in the fissures and constitute what we know as the vein. In many cases the fissures might tend to split and the two portions extend along parallel to each other, separated by blocks or masses of peridotite, and then ultimately converge, and the filling of these separated portions of the fissure with vein material derived from the circulating solution would produce what are known as "horses." In the vicinity where the parallel fissures re-unite, and in the lenticular cavities along the line of faulting, it would be possible for the circulating solutions to interact, producing physical and chemical changes more favorable to the precipitation of the mineral content than in the extremely narrow portions of the fissure where the walls approach each other. Considering the vast extent of this underlying batholith of granodiorite and the long time necessary for it to cool it would be possible for mineral bearing solutions to be given off and to circulate upwards long after the solidification of those portions nearest the surface.

Throughout the whole region involved in this report where the peridotite or serpentine outcrops at the surface there are numerous dikes of this granodiorite ranging all the way from a few feet to several hundred feet in width and varying in texture from that closely approaching an andesite porphyry to that of a diorite porphyry. In some of the larger areas they become almost plutonic in character. In several places in the upper portion of Culver Springs gulch some of the extremely narrow dikes almost grade into quartzitic veins. A careful study was made at the contact of these dikes with the serpentine and in many cases the latter was of the bright green schistose variety or dark green to black, resembling obsidian or broken glass. Specimens of serpentine collected within three or four feet of the contact are often heavily stained with iron oxide and contain secondary silica in the form of opal, along with numerous very fine disseminated crystals of pyrite. An assay made of one of these samples taken from the granite-serpentine contact on the hillside just below King creek yielded a return of .04 of an ounce in gold.

The only other rocks available that might be considered as the source of the vein materials are the diabase dikes or the gabbro intrusions. If the conclusion be correct that the diabase dikes were intruded at a later time than the formation of the ore bodies and this conclusion seems pretty well established from their contact relations in the face of No. 5½ tunnel, then the diabase may be eliminated as a possible source. The gabbro intrusives are probably closely associated if not contemporaneous with the diabase. They were intruded as sills into the Swauk formation, as may be observed north of Blewett in Camas land which is outside the area of this map, and are apparently cut by at least one diabase dike on the ridge forming the divide between Negro and Ingalls creeks. The only ores found associated with these gabbro masses are low grade copper sulphides which do not occur in well defined veins but rather as scattered impregnations in the gabbro mass.

Thus there does not seem to be any well grounded evidence for considering the gabbros as the source of the gold ores in the Blewett district. The only conclusion to be drawn is that the mineral bearing solutions must have been derived from the

—6

granodiorite. The actual occurrence of the ore bodies in this district, the size, shape, and relations of the veins to the country rock, conform very closely to the conditions just outlined.

While no positive statement to this effect can be made, there is a strong suggestion that the gold bearing veins in the Blewett mining district have been derived from heated waters given off from the underlying, slowly cooling granodiorite magma. After circulating through the peridotite masses and having helped to produce the serpentinization of the peridotite, they found their way to the partially deformed and squeezed fissure zones and there under gradually reduced temperatures and pressures precipitated their mineral content. When this process of mineralization was finished there is good reason to suspect that this entire region was still covered over by a tremendous thickness of Pre-Tertiary sedimentary and volcanic rocks, which were extensively metamorphosed by the intrusions just discussed.

From this time on to the beginning of the Eocene the region is supposed to have undergone extensive deformation and erosion, so that a large part of the covering was removed and redeposited as sandstone and shale in the early Eocene fresh water lake beds. Then owing to deformational movements of this portion of the Cascades the region itself was submerged until it became a part of the floor of an early lake and upon it accumulated a thick deposit of sedimentary rocks which in turn were again removed by erosion during a more recent uplift of the region. Hence a large part of the rock originally containing mineral veins has been worn away and a part at least has accumulated as placer quartz and gold in the Swauk formation and in the later Quaternary gravels.

PLACER DEPOSITS.

The placers were the first deposits to be worked in this district and more or less attention has been paid to them ever since. They are confined to the stream valleys, and chiefly to Peshastin and Negro creeks. The richest and most extensive deposits occur north of Blewett on the Peshastin, near the mouth of Negro creek. These gravels were originally derived from the upward extension of the present quartz veins in this district, by the erosive action of the smaller gulches and creeks emptying into Peshastin and Negro creeks. These old channels occupy in a general way the trend of the present valleys but do not always conform to the present winding of these streams. The downward cutting of the streams initiated by the upward movement of the mountain mass carried onwards the rounded pebbles of quartz and country rock and deposited them where they are now found. The placer gravels vary in thickness and character and may be divided into two groups, one representing the gravels as they were originally deposited, and the second those that have been worked over by the present streams. The former comprise the bench gravels and a part of those lying next to the bed-rock; the latter, those gravels on a general level of the present creek beds. In general the older gravels contain the richest deposits of gold. Through these the gold is pretty evenly distributed, but is richest close to bed-rock. Where the bed-rock is composed of slate many crevices and water-worn pot holes contain exceptionally rich deposits of gold. An instance of this may be cited on Peshastin creek, near the mouth of Negro creek, where nuggets having a value of $6.75 are said to have been taken out. On Negro creek, for a distance of two miles above its mouth, wide bars occur which have been worked by sluicing. The average depth of these bars is about five feet and the average pay ranges from ten cents to $1.25 per cubic yard.

CHAPTER IV.
DETAILED DESCRIPTION OF THE MINES.

INTRODUCTION.

In the examination of the mine properties in this district, all of the accessible underground workings were visited, and the ore bodies carefully studied, as well as the character of the wall rock. Unfortunately many of the old workings had caved in and were entirely inaccessible. In such cases it has been necessary to rely upon the older mine surveys and upon such statements as could be obtained from the older mining men who had worked there and whose statements seemed most authentic.

Some of the properties have been extensively developed and others are represented by merely a few feet of surface tunnels. In all there are at the present time about forty claims in the district, owned by fourteen different companies or individuals. Eight of these claims are patented. In addition to these a large number of claims have been abandoned. In describing each of the individual properties the more salient features will be presented in the following order: the geographical location; the history of the development; the underground workings and production; and the economic geology. The latter will include a description of the country rock, the form, distribution, and character of the ledges, and their relation to the country rock.

THE WASHINGTON METEOR MINING COMPANY.

LOCATION. This property consists of seven claims located in Culver gulch and extending from Peshastin creek on the east to its head near the United States mineral monument on the west. The claims are the Culver, Bobtail, Hummingbird, Sandell, Peshastin, Blackjack, and Keynote. The total amount of underground development on this property consists of over 7,000 feet of tunnel in addition to upraises, shafts, and open cuts.

The developments have been made mainly on one vein, which will be known as the Peshastin and which is the same as the Blackjack, Culver, and Hummingbird leads. In addition to the underground development, several miles of aerial tramway, a twenty-stamp mill, cyanide plant, and assay shop have been installed.

HISTORY. The Culver claim was discovered and located by John Shafer in 1874. The Hummingbird by Samuel Culver, the Bobtail by James Lockwood, and the Sandell by John Olden and Peter Wilder, were all located in the same year. A little later all of these claims were bought by James Lockwood and his son, E. W. Lockwood, and H. N. Cooper, who erected a six-stamp mill and operated the property for eight years. It was then sold to Thomas Johnson and again re-sold in 1891 to the Culver Gold Mining Company. A ten-stamp mill was then erected and was connected with the mines by a cable tramway. In 1892 this company was bought by the Blewett Gold Mining Company, who erected the present twenty-stamp mill, situated at the mouth of Culver gulch, and which has been described under the heading, "Treatment of Ores." A steam sawmill was built about three miles above Blewett to supply timber necessary for carrying on mining and milling operations. All of the machinery used in this mill was brought in over the road from Cle Elum. Until 1894 the mill and mine were operated by the same company, but after that year the practice was made of leasing different portions of the mine while the company still operated the mill and charged a royalty on the output, as well as the cost of milling. It was estimated that about sixty men were thus employed continuously for three or four years. About this time it was found that a large part of the gold was not caught on the amalgamating plates, due to the presence of arsenic, but ran off in the slimes. Consequently a cyanide plant was erected under the direction of A. J. Morse, for Rosenberg and Company, one of the lessees, and 600 tons of tailings were treated, giving values ranging from $3 to $30 per ton.

In 1896 the Blewett Mining and Milling Company passed into

the hands of the Warrior General Company, which, in turn, in the year 1901, was bought by the Chelan Mining and Milling Company. In the same year Mr. Thomas A. Parish, the owner of the Peshastin and Keynote claims, organized the La Rica Mining Company and later acquired the Blackjack property. Later both the La Rica and the Chelan companies were consolidated as the Washington Meteor Mining Company. This company has continued to operate to the present time. In 1910 the property was leased for a term of five years to a company composed largely of stockholders and known as the Blewett Mining and Leasing Company. During the past year they have taken out the amalgamating tables from the stamp mill and installed in their place cyanide tanks.

UNDERGROUND WORKINGS. The underground workings of this property may be best seen by referring to Plates VIII and IX. The oldest workings occur near the head of Culver gulch, in the vicinity of what is known as the Summit pocket, and are represented by the various underground workings associated with tunnels 1, 2, 3, and 4, on the west side of the big basaltic dike. The most easterly workings on this property are located on the Blackjack claim, and consist of a tunnel driven westerly on the vein at the level of Peshastin creek for a distance of 1,300 feet, together with shafts and upraises. Three thousand tons of ore are said to have been stoped out, yielding an average value in gold of $10 per ton. Several other similar openings have been made on the vein upon this same claim. Proceeding westward, the next developments are found in the Meteor tunnel, located on the Peshastin claim. This tunnel consists of a diagonal cross-cut to the vein and then a drift westward for about 700 feet. Several upraises were made which connect with the Peshastin tunnel.

The Peshastin tunnel is about eighty feet higher, and consists of a cross-cut tunnel extending from the center of Culver gulch southward for a distance of 300 feet, where it intersects the vein. Drifts have been run both to the east and to the west for a distance of over 1,000 feet, and an upraise driven to connect with the Sandell tunnel.

The Sandell tunnel is a cross-cut extending southward for 300 feet, where it cuts the main Peshastin lead 180 feet higher than the level of the Peshastin tunnel and consists of a series of drifts on the vein.

The Draw tunnel, whose portal is situated in Draw canyon, at an elevation of 120 feet above that of the Meteor tunnel, has been driven into the west directly on the vein, and is connected by upraises from the Meteor and Peshastin tunels below. Several large and very rich ore bodies have been worked from these tunnels.

The next tunnel to the west is the Hummingbird, which consists of a cross-cut from Culver gulch southward for a distance of 150 feet, where the same vein is intersected. Drifts have been run in on this for about 500 feet. Further up the canyon and to the west of this is the Bobtail tunnel, which has been driven directly on the vein. This is locally known as the Y tunnel. On the western end of the Bobtail claim, the tunnel known as No. 9 has been driven in zigzag fashion, resulting from failure to find definitely the position of the vein. As the work progressed this tunnel was extended on into the Culver claim and crossed the south side line of the Culver into the Pole Pick No. 2, at the northeast corner of that claim.

About 300 feet west from the mouth of No. 9 tunnel a cross-cut known as No. 8 tunnel has been driven northward a distance of 200 feet to the vein. At a point about twenty-five feet in No. 9 tunnel, back from the point where the tunnel crosses the side line of Culver claim into the Pole Pick claim, an upraise has been made on the vein for a distance of 266 feet to the Blewett tunnel. This tunnel is a cross-cut from the center of the gulch, northward to the vein and about 100 feet beyond and again reaches the surface by an easterly drift on the vein. At the point where the upraise from No. 9 tunnel reaches the Blewett tunnel a continuation of the upraise has been made up to No. 6 tunnel, which has been driven in, westerly, on the vein. Below No. 6 tunnel, a tunnel known as No. 6½ has been driven westerly on the vein which is connected by an upraise with No.

6, No. 6 is connected with No. 5½ and the latter with No. 5, both of which have been driven on the vein. No. 5½ extends to the diabase dike, where the vein has been cut in two by the intrusion. A large part of the ore from the level of No. 9 tunnel up to No. 5 has been stoped out and transferred to the stamp mill at the mouth of Culver gulch by a bucket tramway. A large part of this work was done by leases from 1894 to 1900.

GEOLOGICAL RELATIONS. The veins in this property represent an east to west zone of fissuring which has been filled with quartz, calcite, and talc, heavily impregnated with arsenopyrite and pyrite, and containing varying amounts of gold. The country rock is serpentine in various stages of alteration. In the vicinity of the veins, there are prominent reddish colored bands, due largely to staining with iron oxide, and containing many small seams of calcite. In places this grades into a greenish rock and has been known locally by the miners as porphyry, but which upon examination is found to be an altered phase of serpentine.

The general trend of this vein is north 75° west, with a very steep pitch to the south. The vein varies in width from a thin seam to sixteen feet. On the Peshastin claim two ore shoots were found known as the "Parish" and the "McCarty" shoots. These were worked mainly from the Peshastin tunnel but they extended down to the Meteor tunnel. They pinched out to a very narrow seam above the level of the Peshastin tunnel and also narrowed down both to the east and to the west along the general direction of the vein. The ore in these shoots was largely calcite and talc with some quartz, but rich in arsenopyrite and pyrite. Specimens exceedingly rich and containing free gold were common. The gold occurs in flakes and in wire-like form and it is especially noticeable in the talc. In the workings to the west of these ore shoots the veins become much narrower, and in some cases are very low grade. The walls in places are ill defined and hard to follow. Still farther west on the Hummingbird and Bobtail claims, they again widen out and rich ore has been obtained. In the upper workings, extending from No. 9 tunnel to the head of the gulch, in the Summit pocket, the ores were exceptionally rich, due largely to their lying in an enriched oxidized zone. Towards the head of the gulch, the main vein has a tendency to split and again coalesce. It is here that some of the richest ore ever taken out of the camp occurred. Some of this is said to have yielded $60,000 to the ton.

PRODUCTION OF THE MINE. No detailed records have been kept of the production of this mine but according to Gen. J. D. McIntire, who operated these properties for several years, the Washington Meteor Company together with those which preceded it have received approximately $1,200,000 in gold.

FUTURE OF THE MINE. Considering the general character of the ore bodies which have already been mined, the relation of these to the country rock, and the theories which have already been discussed relating to the origin of these ore bodies, the conclusion is drawn that lenses of rich ore, with intervening barren stretches of low grade ore, will extend throughout the entire east to west limits of the property, to a depth of at least 1,000 feet below the lowest workings, namely the Blackjack tunnel at the level of Peshastin creek. Whether this can be mined profitably or not depends upon whether the returns from the lenses of very rich ore will be greater than the cost of mining in the intervening stretches of low grade ore. If future operations are carried on intelligently and economically, they should be made to pay.

ALTA VISTA MINE.
(POLE PICK NO. 2 CLAIM.)

This property lies on the south side of Culver gulch about one mile west of Peshastin creek. It joins the Culver claim on the north and the Pole Pick No. 3 of the Cascade Mining Company's group on the west. On the east it is separated by a fraction from Pole Pick No. 1 of the Ellinor Mining Company. The position and underground workings of this claim may be referred to in Plates VIII and IX.

This claim was located in 1874 by John Earnest, and was

later acquired by John Shoudy, but in 1906 was sold to Mr. Thomas A. Parish, who organized the Alta Vista Mining Company.

The development work consists approximately of 790 feet of underground tunnel, and several open cuts upon the surface. The largest tunnel is a continuation of No. 9 tunnel from the northeast corner of the claim westward for a distance of 292 feet. Ninety feet west from this northeast corner, a cross-cut has been driven southward for 200 feet, for the purpose of tapping the downward extension of the Pole Pick vein. On the Pole Pick No. 2 fraction just east of this claim, and about 350 feet in elevation above the level of No. 9 tunnel, three short tunnels have been driven westward on the vein for a distance of about 200 feet each. One hundred feet higher a fourth tunnel has been run on the vein for a distance of 100 feet. The long tunnel extending from the northeast corner westward has been driven on the vein and a raise has been run upward for a distance of thirty feet.

This vein consists of quartz, calcite, and talc, heavily impregnated with arsenopyrite and pyrite, and yields assays in gold ranging from $8 to $200. The vein varies in width from four feet to a mere stringer and has a general strike ranging from north 88° west to due east and west.

The south cross-cut extends 90 feet and cuts a well defined fissure zone with slicken-sided walls and is composed of crushed serpentine filled with a network of quartz and calcite vein material containing a large amount of arsenopyrite and pyrite and yielding assays ranging from 0.04 to 1.6 ounces to the ton. About ten feet north of this zone a similar narrow stringer of calcite parallel to the main zone gave assays of $27 per ton. The width of this mineralized fissure zone averages four feet with a nearly vertical pitch and a strike of north 78° east. The cross-cut has been driven ten feet further through a hard flinty rock composed of serpentine thoroughly impregnated with almost colorless, barren quartz.

The three short tunnels on the Pole Pick No. 2 fraction show a more or less oxidized quartz vein with occasional crystals of pyrite scattered through it cutting up through the serpentine, and having a general strike of north 80° west. The pitch varies from vertical to 75° to the north. The vein is much broken, due to the close proximity of the diabase dike. This dike strikes southwest to northeast and in the vicinity of the veins thins and narrows and has apparently cut the vein off

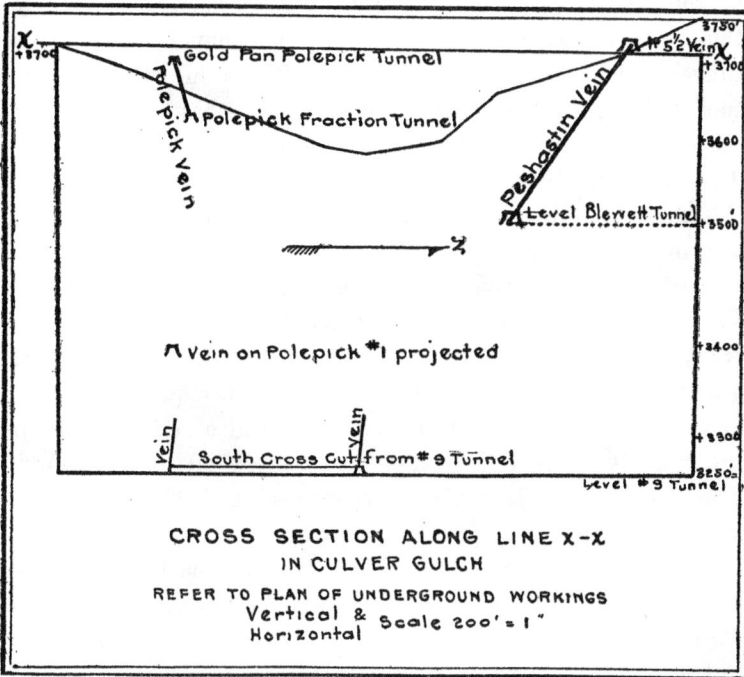

Fig. 1. Cross section to show relation of Polepick and Peshastin veins.

from its connection with the Ellinor Pole Pick No. 1. The Gold Pan-Pole Pick tunnel above the three tunnels on the fraction shows a quartz vein four feet wide and in places is composed of crumbly, iron-stained broken material. One hundred feet from the mouth of the tunnel much calcite enters into the vein. Eight general assays were taken from the vein in this tunnel and none of them gave values lower than $27 per ton.

POLE PICK NO. 1 MINE.

The Pole Pick No. 1 claim lies on the south side of Culver gulch, about one-half mile west of Blewett and is owned by the Ellinor Mining Company. This claim was located in 1883 by Samuel Culver. Later this was bought by Thomas Johnson and on October 16th, 1901, was sold to Chicago parties who obtained a United States patent and now hold it under the name of the Ellinor Mining Company. The development work on this property consists of two cross-cut tunnels driven from Culver gulch, southward to the Pole Pick vein. One of these tunnels is 200 feet long and in it an upraise has been driven on the vein for a distance of 147 feet and at the 100-foot level a drift has been driven west for a distance of 100 feet and from this drift an upraise has been made to the surface. In these workings a larger part of the ore has been stoped out and treated in the ten-stamp mill bought from the Blewett Mining Company. Assays made on this at the time the ore was extracted are said to have averaged from $10 to $132 in free gold. The vein is mainly hard, massive, milky quartz, ranging in width from one to four feet. The trend of this vein is approximately north 80° west and the pitch 80° to the north. The country rock is entirely serpentine. In 1901 it was estimated that this mine had produced about 8,000 tons of ore, having a value of $70,000. Near the surface the Pole Pick lead apparently consists of three veins converging in depth. To the east of this claim the continuation of these veins is hard to trace although certain isolated outcrops may represent them, extending through the Rising Sun claim to the mouth of Culver Springs gulch.

BLINN MINE.

This mine is situated on the western slope of Negro creek and extends from the United States mineral monument at the head of Culver gulch westward to and across Negro creek. This property consists of five claims known as: The Shafer, Olympia, Pole Pick No. 3, Seattle, and Vancouver. They are all patented and owned by the Cascade Mining Company. They were first located in 1878 by Marshall Blinn. In 1889 this property was bonded to Edward Blewett but later passed back into the hands of the Cascade Mining Company.

The development work on this property consists of one long tunnel driven eastward on the vein from Negro creek into the hill, and about 1,000 feet of shorter tunnels distributed over the five claims. On the Shafer claim, near the head of Culver gulch, a tunnel was driven on the vein for the purpose of tracing the ledge into the Peshastin vein. This ledge was found to split up into several smaller ones and become a part of what is known as the Summit pocket, near the mineral monument. In 1880 a two-stamp Huntington mill was erected on Negro creek on the Seattle claim, but on account of the high percentage of sulphides of iron and copper in the ore only a small amount of the gold could be saved. The vein running through these properties has a trend of north 75° west, with a nearly vertical dip and is the western continuation of the Peshastin vein.

On the Olympia claim the country rock is partly serpentine, but the western half of the property, including the Seattle and Vancouver claims, is composed of breccias belonging to the Hawkins formation. The vein varies in thickness from one to five feet and consists mainly of quartz and subordinate amounts of calcite.

NORTH STAR MINE.

This mine is located on the south side and near the upper end of Culver gulch, and is owned by the Golden Eagle Mining & Milling Company. It was first located on October 1st, 1882, as the Golden Phoenix, by William Donnahue, and later sold to Patrick Henry. It joins the Pole Pick No. 1 patented claim on the south. In 1906 it was leased to the Golden Eagle Company, who in 1908 erected a six-stamp mill, with Wilfly tables, etc. It is run by a 25 h. p. Fairbanks-Morse engine. This mill is situated in Culver gulch about one-fourth mile west of Peshastin creek.

While still owned by William Donnahue, a vein of brown oxidized quartz was tapped at a depth of 100 feet by cross-cutting

125 feet. Then levels 100 feet long at intervals of twenty-five feet were run and from these ore was stoped up to the surface. It is reported that 1,000 tons of ore were taken out and treated at the Blewett mill, yielding average returns of $20 per ton in gold. Later, in 1907-08, two tunnels were driven on the vein by Mr. Jack McCarty, the lower one for a distance of 62 feet, and the upper 258 feet. In the lower tunnel the vein is two feet wide and strikes north 70° west, with a vertical pitch. In the upper tunnel, which is about forty feet higher in elevation than the lower tunnel, the strike is north 73° west, and the pitch vertical. At the face of the upper tunnel an upraise has been made on the vein for twenty feet.

The ore is quartz, intermixed with calcite and talc, and heavily mineralized with pyrite. Only a very small amount of arsenopyrite was observed. The vein varies in width from one to eight feet and is enclosed within serpentine walls. On the surface these are stained red by iron oxide and are referred to by the miners as red porphyry. An analysis of this rock has been inserted in the chapter on ore deposits, on page 76. Assays taken at various intervals across this vein range from $1 to $15 per ton, but higher values are said to have been obtained from the Donnahue opening on the same vein during the early days of mining. This is a distinct fissure vein, approximately parallel to the Peshastin and may be assumed to have been formed at the same time and in the same manner. Future development may be expected to show irregular lenses of high grade ore with barren intervals between, lying in the plane of fissuring.

GOLDEN EAGLE MINE.

This claim is situated on the north side of Culver gulch, about one-half mile above its junction with Peshastin creek. It was located in 1902 and is owned by Jack McCarty of Blewett. The development work consists of an upper and lower tunnel. (Diagram on Plate X.) The upper tunnel has been run as a cross-cut in a direction north 25° west, for a distance of 110 feet, to the vein, and from the face a drift has been run on the vein westward for a distance of sixty-three feet. Four feet from the main cross-cut, on the drift, an upraise has been made on the vein to the surface. The lower tunnel represents 650 feet of work and is mainly a cross-cut extending due north with several branches.

The vein is composed of quartz, calcite, and talc, impregnated with iron pyrite, and having a thickness varying from one to three feet. It trends in a direction north 80° east, and pitches about 80° north. The country rock is mainly serpentine, containing isolated blocks of quartzite from the Peshastin formation. When traced to the westward and eastward it is difficult to find any continuation of the lead. The evidence obtained in the field indicates that as the vein extends westward it passes into the Hawkins breccia and becomes narrow and irregular but continues on as a part of the Golden Guinea vein on the Negro creek slope. Two assays taken across the vein from the upper tunnel, near the upraise, gave results of $3.75 and $4.10 per ton.

TIP TOP MINE.

This property consists of two claims, the Tip Top, and Golden Cherry, and is owned by the Tip Top Mining Company. These claims are situated on the east side of Peshastin creek and extend from the town of Blewett eastward toward Windmill point.

The Tip Top claim was located by John Sumners in 1880. It was first worked by the Tip Top Mining Company. A shaft on the ledge was sunk for a distance of seventy-five feet and two cross-cut tunnels 380 and 400 feet in length were driven into the vein. The ore was stoped out to the surface and run through an arrastra. This oxidized ore is said to have given an average return of $40 per ton. In 1888 it was abandoned and again re-located in 1889 by T. J. Vinton, who in 1895 leased it to several parties, who took out ore and treated it at the Blewett mill. All of the tunnels on this property were inaccessible at the time of examination and most of the information was obtained from miners who had previously worked there.

The vein is said to have averaged two and one-half feet in width, and to have had a general east-west strike, with a pitch to the north. The country rock is partly serpentine and partly breccias of the Hawkins formation.

The Golden Cherry claim, which was originally known as the Golden Chariot, was located about 1880. Later it was known as the Eureka and now as the Golden Cherry. Up to the present time about 970 feet of underground work has been done. The portal to the main tunnel is situated on the level of Peshastin creek about 2,000 feet south of Blewett and below the county road. The country rock is serpentine, but much fractured. Some ore was found in the south tunnel. The assessment work for the Tip Top group is now being done entirely in the Golden Cherry tunnel.

WILDER MINE.

This group consists of three claims known as the Ivanhoe, Kennilworth, and Amber Glee, situated on the north side of Culver gulch and over the divide down towards Negro creek. The Ivanhoe was originally known as the Hindoo, later as the White Elephant, and was first opened in 1881. It was re-located as the Ivanhoe in 1902 by James Wilder. The Kennilworth, which joins it on the north, and the Amber Glee, which is an extension on the east, were also located in the same year. The underground workings on the Ivanhoe consist of two tunnels driven in on the vein, one of which is 140 feet long and the other 160 feet, with a forty-foot cross-cut. This work was done in the years 1902 and 1906. The Kennilworth has one tunnel forty feet in length. The Amber Glee has one cross-cut tunnel driven north to the vein with a drift of 200 feet and an upraise of sixty feet.

The width of the vein varies from two to six feet, carrying gold and a small amount of silver. The trend of the vein is nearly east and west, with a nearly vertical pitch. It is estimated that 150 tons of ore have been taken out, and assays made by Mr. Bogardus in 1906 yielded returns ranging from a trace to $72 per ton in gold with a very small amount of copper and silver.

GOLDEN GUINEA MINE.

This property consists of eight claims known as the Golden Guinea; Golden Guinea No. 1, east; Golden Guinea No. 1, south; Golden Guinea No. 2, south; the Brown claim; the Lillian; and the Copper Queen. They were first located in 1897 and were known as the Francis L. group. Later they were relocated in 1904 by R. F. Brown, of Leavenworth, and were named the Golden Guinea group. These claims are situated on the north and south slopes of Negro creek, about two miles above its mouth. Development work consists of 233 feet of tunnel on the Golden Guinea claim, with a shaft forty-three feet deep. On the Golden Guinea No. 1 south, a cross-cut tunnel has been driven south for a distance of fifty feet for the purpose of tapping the east-west vein. Several smaller tunnels and open cuts have been made on all of the claims. The Golden Guinea vein lies in the Hawkins formation and has a general trend of north 70° west. It is represented by a belt ranging from ten to fifty feet in width, composed of iron stained country rock, containing more or less discontinuous stringers of rusty, broken quartz, occupying a zone about two feet in width. When traced eastward this vein apparently is headed for the Golden Eagle vein in Culver gulch. No deep workings exist so that it is impossible at the present time to predict what future developments may show. Several assays were taken on these veins but none of them ran over $1 per ton in gold, although much higher returns are claimed to have been previously obtained. This property is owned by R. F. Brown and F. W. Losecamp, of Leavenworth, Washington.

LUCKY QUEEN MINE.

The Lucky Queen group consisting of the Lucky Queen and Bee Queen claims, is located a short distance north of Blewett on the east side of Peshastin creek, and adjoins the Blue Bell claim on the north. This claim was originally located by John

—7

PLAN OF UNDERGROUND WORKINGS OF THE GOLDEN EAGLE

E   F

E   Cross Section   F

PLAN OF UNDERGROUND WORKINGS OF THE LUCKY QUEEN

G   H

N

G   Cross Section   H

Scale 200 ft. = 1 in.

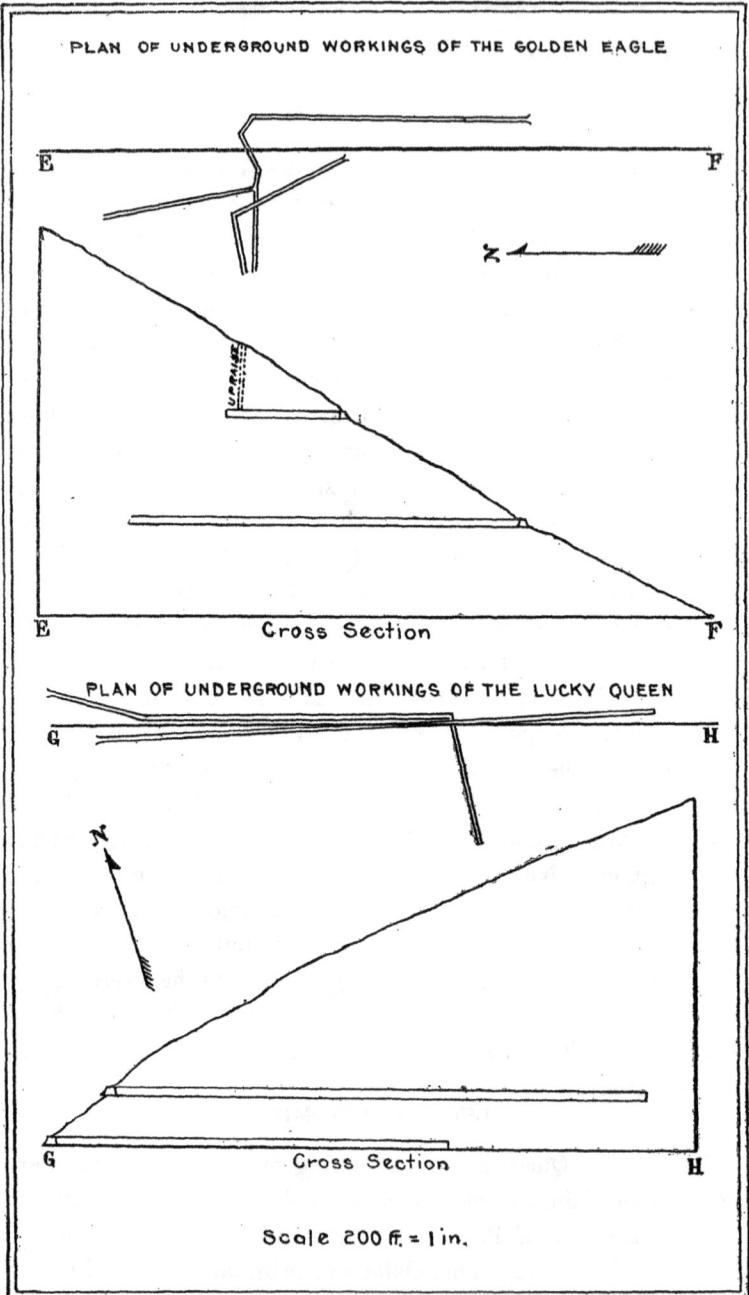

Plan of underground workings and cross-sections of Lucky Queen
and Golden Eagle mines.

Ernest in 1889 but later passed into the hands of Peter Anderson and Thaddeus Neubaur in 1895.   The development work on this property consists of two tunnels, one driven at the level of the county road and the other about forty feet higher.   Altogether about 1,000 feet of tunnel have been driven.   The lower tunnel extends in 425 feet from the mouth, and at the face, a cross-cut turns off at right angles for a distance of 120 feet. The vein is parallel to the main drift, composed of talc, calcite, and quartz, and varies in width from a mere stringer to three feet.   In the upper tunnel a drift has been run in on the vein for a distance of 412 feet, which strikes north 78° west and pitches nearly vertical.

The ore in this vein is similar to that below and is known to have yielded rich specimens showing free gold.

BLUE BELL MINE.

The Blue Bell mine is situated one-quarter mile north of Blewett, on the east side of Peshastin creek, near the county road. This was located by John Bomaster in 1890, and was formerly known as the "I. X. L."   The development work consists of about 450 feet of tunnel driven through a badly fractured and dislocated mass of serpentine.   No well defined vein could be found.   Many of the seams in the country rock are filled with vein material composed of quartz and calcite, with pyrite and some gold, but the strike and pitch varies from point to point. In some very small pockets, high values are reported to have been taken out by Mr. Bomaster, but none were seen by the writer at the time of examination.   Eighty-five feet from the mouth of the tunnel the vein splits, the north wall is composed of iron stained serpentine, with many seams of calcite, and the south wall of ordinary green serpentine.

PROSPECT MINE.

This property is situated west of Blewett on the north slope of the ridge leading down to Culver Springs gulch, and extends from its head down nearly to Peshastin creek.   It consists of a group of five claims known as the Sunset, the Sunset Extension, the Redjacket, the Lone Star, and the Katy, all of which are owned by the Prospect Mining & Milling Company. Openings have been made upon all of these claims and some of the ore was taken out in early days and treated in an arrastra. The development work on this property consists almost entirely of the required assessment work.

All of the workings on this property are situated in the serpentine formations and the vein material consists mainly of quartz, with calcite and talc, containing gold and a little silver. All of the openings are near the surface and consequently in the oxidized zone. The vein can be traced on the surface by the red iron-stained character of the serpentine and occasional outcrops of quartz.

HOMESTAKE MINE.

This claim lies on the west side of Peshastin creek, near the mouth of Culver Springs gulch. One hundred sixty-five feet of tunnel have been driven in on a quartz vein trending about north 75° west, and pitching nearly vertical. The wall rocks are iron stained serpentine, containing calcite seams. A general assay taken on this vein from the mouth of the face of the tunnel gave returns of 0.04 ounce in gold and 0.2 ounce in silver. It is possible that this vein represents the eastward extension of the Pole Pick. This property is owned by Mr. John Olden, of Blewett.

LONE ROCK PROSPECT.

This claim is situated on the east side of Peshastin creek just back of Blewett, on the slope of the hill leading up to Windmill point. No ore has ever been taken out and the development work has consisted entirely of assessment work. A tunnel eighty-eight feet in length has been driven on a narrow belt of red, iron-stained serpentine seamed with small stringers of calcite. A general assay of this rock was taken, which yielded returns of 0.02 ounces of gold and 0.2 ounces silver. This claim is owned by Mr. John Olden, of Blewett.

JOHNSON PROPERTY.

The Johnson property consists of three claims extending in a general east-west direction, about three-quarters of a mile north of Blewett, and extends from Peshastin creek easterly to the summit of the ridge leading down to Ruby creek. These three claims are known as the April Fool, the Venus, and the Donaldson.

The development work on this property consists of a tunnel about 100 feet long on the April Fool claim, driven on a vein pitching nearly vertical. On the Venus claim a tunnel has been driven in nearly 115 feet in a direction north 20° west. No vein was found here, however. On the Donaldson claim a tunnel was driven in for a distance of 195 feet directly east. This tunnel, however, was inaccessible at the time of examination, but is reported by Mr. Johnson to have encountered a diabase dike at a point sixty feet from the mouth. Beyond that point the Swauk conglomerate constitutes the south wall. A second tunnel has been driven on this same claim along the direction north 40° east for a distance of eighty feet. At the top of the ridge looking down into Ruby creek an open cut has been made on a four-foot vein composed of talc, with serpentine for both walls. About twenty-five feet to the south of this cut the basal beds of the Swauk formation outcrop. Assays made on this ore gave returns of 0.02 ounces gold and no silver. However, much higher assays are said to have been obtained from some of the ore extracted. These veins are not distinct but much broken and fractured. They cut through the Peshastin and peridotite formations and lie close to the contact with the Swauk sandstone.

INDEX

PUBLICATIONS

OF THE

WASHINGTON GEOLOGICAL SURVEY.

Volume 1.—Annual Report for 1901. Part 1, Creation of the State Geological Survey, and An Outline of the Geology of Washington, by Henry Landes; part 2, The Metalliferous Resources of Washington, except Iron, by Henry Landes, William S. Thyng, D. A. Lyon and Milnor Roberts; part 3, The Non-Metalliferous Resources of Washington, except Coal, by Henry Landes; part 4, The Iron Ores of Washington, by S. Shedd, and the Coal Deposits of Washington, by Henry Landes; part 5, The Water Resources of Washington, by H. G. Byers, C. A. Ruddy and R. E. Heine; part 6, Bibliography of the Literature Referring to the Geology of Washington, by Ralph Arnold. Postage 20 cents. Address, State Geologist, University Station, Seattle, Washington.

Volume 2.—Annual Report for 1902. Part 1, The Building and Ornamental Stones of Washington, by S. Shedd; part 2, The Coal Deposits of Washington, by Henry Landes and C. A. Ruddy. Postage 20 cents. Address, State Librarian, Olympia, Washington.

Bulletin 1.—Geology and Ore Deposits of Republic Mining District, by Joseph B. Umpleby. Bound in cloth; price, 35 cents. Address, State Librarian, Olympia, Washington.

Bulletin 2.—The Road Materials of Washington, by Henry Landes. Bound in cloth; price, 60 cents. Address, State Librarian, Olympia, Washington.

Bulletin 3.—The Coal Fields of King County, by George W. Evans. In preparation.

Bulletin 4.—The Cement Materials of Washington, by S. Shedd and A. A. Hammer. In preparation.

Bulletin 5.—Geology and Ore Deposits of the Myers Creek and Oroville-Nighthawk Districts, by Joseph B. Umpleby. In press.

Bulletin 6.—Geology and Ore Deposits of the Blewett Mining District, by Charles E. Weaver. Bound in cloth; price, 50 cents. Address, State Librarian, Olympia, Washington.

Bulletin 7.—Geology and Ore Deposits of the Index Mining District, by Charles E. Weaver. In press.

Bulletin 8.—Analyses and Fuel Values of Coal from Washington Mines, by George W. Evans. In preparation.

Bulletin 9.—The Underground Water of the Coal Fields of Washington. In preparation.

(OVER)

PUBLICATIONS OF THE U. S. GEOLOGICAL SURVEY, IN CO-OPERATION WITH THE WASHINGTON GEOLOGICAL SURVEY.

(For copies of these publications address the Director, U. S. Geological Survey, Washington, D. C.)

Topographic Maps of the Following Quadrangles: Mount Vernon, Quincy, Winchester, Moses Lake, Beverly and Red Rock. Price, 5 cents each.

Water Supply Paper No. 253: Water Powers of the Cascade Range, Part I., Southern Washington.

Water Supply Paper No. —: Water Powers of the Cascade Range, Part II. In preparation.

Water Supply Paper No. 272: Results of stream gaging in Washington for 1909.

PUBLICATIONS OF THE U. S. DEPARTMENT OF AGRICULTURE, BUREAU OF SOILS, IN CO-OPERATION WITH THE WASHINGTON GEOLOGICAL SURVEY.

(For copies of these publications address one of the members of Congress from Washington.)

Reconnoissance Soil Survey of the Eastern Part of the Puget Sound Basin, Washington.

Reconnoissance Soil Survey of the Western and Southern Parts of the Puget Sound Basin, Washington. In press.